Sports and the Env

Conflicts and So

– a Manual

This study has been performed on behalf of the German Federal Environmental Agency under the environmental research program of the German Federal Ministry for the Environment, Nature Conservation and Nuclear Safety (Project no. 295 11 140). The opinions expressed in this publication are not necessarily those of the publisher.

The complete version of this work (719 p.), by Schemel and Erbguth, 2000 (which also treats related legal questions in Germany) has the title, "Manual on Sports and the Environment: Objectives, Analyses, Evaluations, Solutions, and Legal Aspects" (available in German only).

Sports and the Environment

Conflicts and Solutions
- a Manual -

Abridgement
of the third completely revised edition

by
Dr. Hans-Joachim Schemel
Büro für Umweltforschung und Umweltplanung, Munich, Germany

Translated by
Arne A. Jaaska,
in consultation with
Wolfgang Strasdas

Published by
the German Federal Minister for the Environment,
Nature Conservation and Nuclear Safety (BMU),
with the support of the German Federal Agency of the Environment (UBA),

German Sports Confederation (DSB),

German Conservation Association (DNR).

Meyer & Meyer Sport

Original title:
Handbuch Sport und Umwelt (by H.-J. Schemel + W. Erbguth)
3. überarbeitete Auflage, ungekürzte Fassung
- Aachen: Meyer und Meyer Verlag, 2000
Translated by Arne A. Jaaska, Wolfgang Strasdas

British Library Cataloguing in Publication Data
A catalogue for this book is available from the British Library

Schemel, Hans-Joachim:
Sports and the Environment: Conflicts and Solutions / Hans-Joachim Schemel.
[Transl.: Arne A. Jaaska, Wolfgang Strasdas].
- Oxford : Meyer & Meyer Sport (UK) Ltd., 2001
ISBN 1-84126-051-7

Oxford, Aachen, Olten (CH), Vienna, Québec,
Lansing/Michigan, Adelaide, Auckland, Johannesburg, Budapest
Member of the World
Sportpublishers' Association

Photos: Volker Minkus, Foto Design Agentur, Isernhagen 51, 57;
Sportpressefoto Bongarts, Hamburg 62, 71, 105;
M. Scheuermann 80, 87, 163; Werner Ernst 99; ADAC Motorsport/Archiv 116;
Akademische Fliegergruppe a. d. Universität Karlruhe e. V. 122; R. Richard 128;
Kanuverband 134; Niessen 141; Dieter Kattenbeck/Frankenboot 151; K. Schrag 177
Cover design: Birgit Engelen, Stolberg
Cover and type exposure: frw, Reiner Wahlen, Aachen
Editorial: John Coghlan
Printed and bound in Germany
by Druckpunkt Offset GmbH, Bergheim
ISBN 1-84126-051-7
e-mail: verlag@meyer-meyer-sports.com

Inside pages printed on 100 percent environmentally-friendly recycled paper.
Cover printed on chlorine-free bleached material.

Table of Contents

Foreword

Since the publication of the first edition of SPORT UND UMWELT (Sports and the Environment), the manual has proved an indispensable aid for anyone concerned with the environmental aspects of sporting activities. We are delighted that the success of the German-language version has now made it possible to publish a shortened version in English.

This book offers solutions to the diverse environmental problems which stem from the extremely dynamic development of sport. It gives a lucid and competent analysis of the problems, with the focus always on offering solutions. This book is especially valuable as a neutral source of information in discussions between nature conservationists and sports representatives concerned to reach sensible and responsible solutions to conflicts of interest. The manual therefore supports the long-standing approach of co-operation and trust with regard to nature conservation which the German Sports Confederation (Deutsche Sportbund) has adopted with deep conviction and considerable success. This book will help to ensure that the future of sport is based on the principle of sustainability. There can be no alternative.

In conjunction with the co-publishers – the German Federal Ministry for the Environment (Umweltministerium der Bundesrepublik Deutschland) and the German League for Nature and Environment (Deutscher Naturschutzring) – we hope this manual has a positive influence on the sustainable development of sport and leads to definite solutions.

Dr. Hans-Georg Moldenhauer, Vice-president of the German Sports Confederation (Deutscher Sportbund)

I. GENERAL SECTION

1. Introduction

The relationship between sport and the environment is often fraught with tension. Sports enthusiasts use the land and its natural resources. What holds true for every other use of land also holds true for sport and recreation: the environment can bear only a limited amount of disturbance and scarce resources must be carefully used, if we are to ensure a future beneficial to mankind.

Though in comparison to other causes of environmental destruction (agriculture, industry, urban growth, and traffic), sport and recreation play a relatively minor role, their contribution to the problem should not be underestimated. Many valuable opportunities exist for engaging in sports in a manner compatible with the principles of sustainability. The selection of a site for sports facilities, the design of these facilities, the choice of a mode of transportation, the choice of materials, and – when sports activities are to take place in nature – the selection of an area ecologically compatible with the activities (one must differentiate here between sites that can tolerate stress and those particularly susceptible to environmental disturbance), all present opportunities for this. In this way we can avoid placing serious stress on nature and the environment without calling the validity of sport into question.

The present handbook hopes to offer a base of information for a sound and fair manner of dealing with frequently recurring conflicts between sport and recreation on the one hand and the environment on the other. It intends to show the possibilities for a responsible exercise of sport that safeguards natural resources. It addresses interested lay people as well as professionals in sports organizations, environmental and nature protection groups, local government, conservation and sports agencies, science, the media and sport-related tourism.
The points of contact between sport and the environment are various and many-sided. They do not admit of hastily formed judgement. In order to study these points of contact in more detail, we must carefully consider the following:

- Type of activity: what kind of sport is involved?
- Behavior: how is the sport practiced?
- Numbers: how many players and spectators are involved?
- Time: at what time of the day or season of the year do the activities take place?

- Site: where are the sports facilities to be built or the activities to take place?
- Infrastructure: what other buildings and ancillary facilities are involved?
- Operation of the facility: what effects do the maintenance and operation of facilities entail?
- Resource consumption and traffic: what potential exists for saving energy and water, reducing waste, and providing environmentally sound forms of transportation?
- Secondary effects: what consequences with regard to infrastructure, traffic, and construction are to be expected outside the facility?

The environmentally friendly exercise of sport is part of the attempt to bring the life of humans on our planet in accordance with the **principle of sustainability**. Since the United Nations Conference on Environment and Development in Rio de Janeiro in 1992 this principle has been internationally recognized – at least on paper. It stipulates that natural resources should be protected from over-exploitation so that they may be available to future generations in sufficient quantity and quality.

In order to reach this goal the demands of the environment must be understood and presented as an integral part of other policy areas. Sport is such an area, as many sports representatives and officials recognize. Fairness, an important maxim among sports aficionados, and the desire to practice sports activities in a healthy environment have given many sports enthusiasts a concern for nature and the environment.

"Sustainable development" is the descriptive term for a type of "development in which the needs of present generations should be met without endangering the needs of future generations. The recognition that environmental problems cannot be treated in isolation from economic and political development, but require a unified approach to problem-solving, stands as a corollary to this term" (Wuppertal-Institut, 1996). Thus the present understanding of the term "sustainability" comprises both an ecological dimension (the use of natural resources according to their ability to be renewed) and a socio-economic dimension: human activities within a given area must ensure an acceptable economic and social standard of living over the long term.
When transferred to the field of sport and recreation, this principle means that sport in and of itself is not called into question. Sport fulfills very important social functions, promoting for example, a sense of well-being, developing the

personality, maintaining health and providing a release for occupational stress. Yet sport is connected in turn to a sound environment and to a landscape offering qualities such as closeness to a natural state, diversity, uniqueness, and beauty.

The following **"rules for sustainable management"** in conjunction with the German Bundestag's task force report "Protection of Man and the Environment" (Schutz des Menschen und der Umwelt) (1997) may be regarded as the most important conditions for a long-term "system sustainability", or assurance that future human needs will be met:

- The use of renewable natural resources (for example, forests or stocks of fish) may not over the long term exceed their rate of regeneration – otherwise these resources will be lost to future generations.
- The use of non-renewable natural resources (for example, fossil fuels or agricultural land) may not over the long term be greater than substitutions for their functions (example: a possible substitution for fossil fuels would be hydrogen from solar electrolysis).
- The use of irreplaceable natural and cultural treasures (for example, endangered animal and plant species as well as architectural remains of the past) may not endanger the qualities that make them worthy of protection – otherwise these resources will be irretrievably lost. (e.g. respecting limitations placed on use of areas subject to a high degree of protection.)
- Emissions of chemical substances and energy may not over the long term exceed the ability of the natural environment to absorb them (e.g. accumulation of greenhouse gases in the atmosphere or of acidic substances in forest soils).

These rules for sustainable management affect only some aspects of sport and recreation. Above all they require us to use non-renewable sources of energy sparingly, to avoid emissions of air pollutants, and to practice outdoor sports in an environmentally considerate way.

The following means for achieving sustainable development are available (UBA, 1997):

- Planning and legislative instruments
- Economic instruments
- Educational instruments that aim at building up environmental consciousness.

These types of instruments can be applied in various combinations, for example, by:

- Prohibiting construction or an intensification of existing land uses in ecologically valuable zones (planning and legislative instruments).
- Using incentive programs to direct agriculture in these zones toward a more extensive ecologically desirable form of cultivation (economic instruments).
- Aiming information on appropriate behavior in environmentally sensitive areas at those seeking recreation in such areas (educational instruments).

2. Developments in the Field of Sport and Recreation

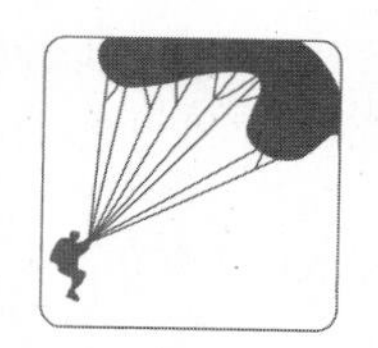

The following socio-economic factors have led to a marked increase in participation in sport and recreational activities:

- Demographic structure: increase in sport as a leisure activity in spite of a stagnant total population (high birth rates in the postwar years, increased interest among women and older people).
- Rising income: participation in sport activities increases with income.
- Increased level of education: a higher level of education works as a strong impulse for participation in numerous leisure activities (including sports).
- Increased opportunities: the infrastructure, technology, and planning required for participation in sport have improved as has the mobility of the population (sport facilities, establishment of recreational areas, availability of transportation, etc.).
- Changed living and work environments: constant reduction in working hours, sport as compensation for increasingly sedentary lifestyles and work places, a growing sense of discomfort and alienation in urban living environments (above all as a result of motorized traffic), as well as increased body and health consciousness.

Next to these quantitative aspects of development one must consider the qualitative changes in people's understanding of sport and their motives for participating. The **change in values** observable in the last 20 years has also left a pronounced mark in the area of sport and recreation.

A key concept in all of this is the "experience-seeking culture" described by Schulze (1992). Against traditional values like postponed gratification, community ethos, subordination (acceptance of hierarchy) and performance, new values have risen that can be characterized by such catchwords as "creativity," "self-actualization" and "identity." The author finds an "increased importance given to pleasure" along with a "new lack of relationships." Transferred to the area of sport and recreation, these tendencies have combined to **decrease** the degree to which people engage in sport as an **organized group activity** and to greatly **increase the variety of sports activities.** In recent times numerous new forms of sports activities have come about, sometimes called "trend sports." New types of experience and physical activity are taken up into the category of sport and recreation.

The new trends in leisure activities are characterized by:

- Individualization/flexibility
- The individual's desire to determine the time of the activity and plan it him-/herself (freedom from constraints)
- Self-experience, self-actualization as a most important personal goal
- Search for new experience and adventure, maximizing experience
- Orientation towards nature, authentic experience of nature
- Emphasis on aesthetics (outfits, sport as lifestyle)
- Commercialization, orientation towards consumption.

What effects do these trends have on the relationship between sport and the environment?

Individualization, meaning the turning away from sports organizations, is noteworthy in that it decreases the opportunities available to influence sports enthusiasts on matters of environmental concern. For sports organizations primarily reach their own members with environmentally conscious policies. Sports organizations can, however, also exercise influence on sports enthusiasts who are not members of an organization and on manufacturers of sporting gear and equipment. Through providing information, education, and formulating codes of conduct (even to the point of stigmatizing certain behaviors in the public mind), sports organizations can sensitize a large part of the population to the needs of the environment. This presupposes a relatively strong readiness on the part of sports associations to extend their influence beyond their membership.

With regard to the behavior of sports enthusiasts, the strong orientation toward a **"landscape to be experienced"** (authentic experience of nature) leads **to consequences** for the environment that merit special attention:

The **increasing desire for "recreation in nature"** should in many cases be interpreted as flight from the high level of environmental stress and lack of green space in urbanized areas. Growing mobility increases the stress upon the landscape, which by means of the automobile is explored into its furthest reaches. Thus the cities have a need for more outdoor spaces suited for recreation that are located near the places where people live. In this way various kinds of recreational activities could be "tied" to their source areas, and traffic that is damaging to the environment could be avoided (see Schemel/Strasdas, 1998 for more details). The desire for variety and a heightened health consciousness also play a role in people's desire for more recreation in nature.

Expansion of the time used for recreation: People have overcome limits once seen to curtail outdoor recreation (extreme temperatures, bad weather), expanded their vacation season into the early part of the year and into fall (as well as "long weekends" and "second vacations"), and made use of new recreational gear and sport clothing (diving and wet suits, for example). The result is an increase in the numbers of people seeking recreation in the outdoors during times when calm and quiet previously reigned supreme in nature.

Expansion of space desirable for recreation: People have pushed into areas that previously "resisted" man, such as steep rock walls (rock climbers), deep water (divers), flat shorelines (windsurfers), rushing streams (canoeists), untouched gorges (canyoning), distant snowfields (tour skiers), the air (hang gliding and paragliding), deep forest (orienteering), or a quiet spot on a riverbank (fishers). The "challenge of nature" is eagerly accepted everywhere, the attractions of nature everywhere enjoyed, even when the plant and animal communities found in the affected areas require freedom from outside disturbance.

Increased **potential for conflict**: The growing interest in nature-oriented recreation runs up against ever-scarcer natural areas. The high demand for recreation in areas close to nature, even in nature reserves and national parks, threatens to overwhelm the ability of these areas to absorb outside stress. This brings calls for restrictions from conservation groups and authorities who commit themselves to defending those areas from increased use. The result is that conservation advocates reap the harvest of incomprehension and anger over regulations-experienced, moreover, during leisure time. Those causing such natural areas to become scarce in the first place (for example, agriculture and road construction) receive less blame.

Effects of environmental degradation on sports participants: The health and mobility of sports enthusiasts is compromised by air and water pollution, but also by the lack of urban parkland and the decreasing availability of attractive open space.

Readiness to act: People's increased consciousness of health and environmental matters leads on the one hand to increased participation in sport and recreation (partly to compensate for sedentary workplaces). On the other hand, it also brings about an increased willingness to conserve natural resources and to reduce environmental degradation as far as possible (including the areas for sport).

3. The Sport-Environment Connection

Since the subject of sport and the environment has become a topic of intense discussion and the search for solutions to conflicts has quickened, sports associations on both the national and international level have taken a stand on environmental questions.
In July 1995 the International Olympic Committee (IOC) held the first "World Conference on Sport and the Environment." At this conference IOC President Samaranch declared the "protection of the environment" to be the third pillar of the Olympic movement (next to "sport" and "culture"). The guidelines used to decide where the Olympic games are to be held were expanded to include environmental criteria. At the end of 1995 a panel of experts was appointed by the IOC to deal with environmental questions. A resolution was passed at the second IOC "World Conference on Sport and the Environment" in 1997 that took the following positions as its starting point:

- "There is a natural partnership between sport and the environment.
- The health and safety of athletes and the sports community depends on an intact environment.
- Sports activities and sporting events should be organized so that not only damage to the environment is minimized, but more importantly, so that both the environment and society benefit, and our environmental heritage is preserved.
- There is a strong correlation between a low level of sports activity and poor health.
- Sport, society, and industry share a responsibility to act in a manner that does not harm the environment and that upholds the principles of sustainability.

It remains to be seen if these first steps and declarations of intent translate into serious measures that pay heed to ecological criteria and limits at international sports competitions.

A clear idea of how important the German Sports Federation (DSB) considers the **responsibility of sports organizations to educate the public about environmental issues** may be had from the following statement by the director of the Federation's "Section for Environment and Sports Facilities." "The basic knowledge that the majority of those active in sport, sports organizations and agencies have on environmental issues is sketchy. Yet these are the people who will have to formulate policy for the future. Generally, people carry over opinions and beliefs on environmental issues that they have learned in their private lives or from their jobs into the area of sport. They have not considered carefully enough whether these attitudes and beliefs apply to the field of sport, or to the specific interests of sport. As in the general population, the base of information on environmental issues of those in the sports community is in the main minimal and knowledge of the ecological side of issues is rudimentary. This indicates that sports organizations have before them a gigantic task to educate the public, and that knowledge of environmental issues must also be incorporated into the various forms of sports education. We have to come to see that self-satisfaction, and looking the other way, will not bring us further, that it jeopardizes the future of sport when the necessary changes in thinking are hindered, watered down or postponed to some indefinite day in the future. The key question is not "What can I do so that everything can stay as it was?" but "What is necessary to provide for the sport of the future?" (Jägemann, 1996).

DSB President Manfred von Richthofen outlined the following position on sustainable sports development in his speech to the conference, "Guiding Principles for Sports Compatible with Nature and the Landscape," held on October 13, 1996 in Wiesbaden. "We do not want to overburden nature with our activities, nor may we do so. This would be counter-productive in many ways. First, the political will, expressed in laws and regulations, to improve the protection of the environment would stand against any such activity. Secondly, we cannot damage or degrade today those places where the sports of tomorrow will take place. After all, we are responsible for the recreational opportunities of future generations. Thirdly, sport, which is increasingly undertaken to improve health and well-being, cannot be presented to the public when it is carried out to the detriment of the environment and natural beauty."

The following forms of environmental stress are to be seen as the **main points where sport has a negative impact on the environment:**
Stress that arises from the **concentration of crowds** (both players and spectators) within a specific area at a specific time. Normally this is connected with a considerable amount of infrastructure and a high level of traffic. When such concentrations occur in relatively sensitive natural areas (for example, in the

mountains in the case of skiing and in shoreline zones in the case of water sports) then particularly grave environmental damage is likely to result. Such stresses can extend to all environmental media, though they are limited to relatively small zones (see chapter C in the special section of this manual).

Stresses arise also from the spread of sports activities into previously undisturbed zones, with respect both to the times and areas in which they take place. In particular the individualization of sports activities (avoiding of crowds), which at first sight seems to be "in harmony with nature", can lead to serious disturbances when such activities take place in sensitive biotic communities. These activities, with their concomitant noise and movement, can displace endangered animal populations or even destroy them when these communities have no other spaces into which they can retreat. The links connecting these first and second types of environmental stress can be quite smooth, for example when sports enthusiasts push far out into the environment from certain concentration points (as when windsurfers set out from marinas in search of remote areas of shallow water or skiers leave ski runs to seek out "untouched" areas).

Noise in urbanized areas (see chapter B in the special section of this handbook) caused by sports activities is a problem (which can be corrected through regulation) only when poor planning has left insufficient space between sports facilities and residential districts. (The special problem posed by motorized sports is excluded here.) Ecological stresses that arise from sports activities in natural areas pose, however, a much greater problem, for they often take place in the few remaining spaces where endangered animal and plant communities live.

A further conflict arises from the **consumption of natural resources** (such as fossil fuels) and the environmental stresses this entails. Resources are consumed, for example, in the manufacture and disposal of sports clothing, in the construction and operation of sports facilities, and in the manufacture and use of vehicles which sports enthusiasts drive to reach the site of their activities.

Automobile traffic is one of the main problems in the area of sport and the environment. The environmental damage caused by driving to and from the site of a given sports activity is indeed among the most serious.

This handbook will, however, only treat them in passing, since the problems posed by automobile traffic and the consumption of resources (energy, water, etc.) are hardly specific to sports activities. Chapter A in the special section goes into more detail on resource consumption and automobile traffic.

The **main points where sports and environmental organizations can work together** lie in the following areas:

Influencing individual behavior: Sports organizations can bring about a more responsible attitude toward environmental resources and the quality of the landscape by educating sports enthusiasts about ecological issues and by sensitizing them to the value of undisturbed nature. In addition to their own members, sports organizations can to some extent also reach sports enthusiasts who do not belong to an organization (for example, through information brochures, public seminars and events, and by their own good behavior).

A heightened environmental awareness and, even more, an increased readiness to bring one's own behavior into line with certain necessary ecological principles are the foundations for the practice of sport that is in harmony with the environment. A decision to forgo certain types of sport activities (for example, placing limitations on when and where one may engage in an activity) thus comes about from the individual's own agreement with the necessity for sensible rules to govern interaction with nature and the environment. He or she does not regard it then as regulation "from on high."

Participation in land-use planning: Planning at the regional and local level sets important parameters on the behavior of individual sports participants. Zoning for recreational areas to have varied levels of use and decisions as to which land to declare as nature reserves provide representatives of sports and environmental agencies with many good opportunities for working together. A conflict between sport-related noise and residential use of land can be avoided when, for example, sports organizations that would be affected by the construction of housing near an existing sports facility work early on to oppose it. Further important opportunities for working together exist in the planning of roadways (or in the decision not to build them), in channeling and managing visitors to recreation areas, and in locating and constructing sports facilities in an environmentally sound way.

Influencing business: Manufacturers of sports gear and articles play a determining role in promoting and increasing the popularity of trend sports, which continually place demands on natural resources. Established sports, such as skiing and motorsports, also have close ties to commercial interests.

From the businessman's point of view growth is always desirable. The environmentally concerned sports enthusiast, however, realizes that limits have to be set on growth when this threatens nature and the environment.

Advertising offers sports organizations a way to exercise a corrective and restraining influence. When, for instance, manufacturers use advertising to promote the unrestrained use of their product in a natural area (or promise adventure with the product that would entail negative effects on the environment), a protest with clearly formulated goals can bring an end to such activity.

Taking an active role in the political process: The state of the environment is affected not only by political action that focuses primarily on the environment, but also by legislation governing traffic, construction, agriculture, business and finance to name only a few. All of these have a profound effect on the environment and so on the acknowledged interests of the sports community. Sports organizations (as the largest "citizens action group") must be encouraged to get involved in all areas of the political process where sports have influence. There they must push for the protection of the environment and the principle of sustainability in the use of natural resources. By working together with environmental and nature preservation groups, sports organizations can play a role in ensuring that the future of our civilization is not jeopardized by shortsighted special interest politics.

4. Handling Conflicts Between Sport and the Environment

Environmental groups and authorities have the protection of nature and the environment as their priority and accept sports activities so long as these do not impinge on this value. Sports enthusiasts and organizations on the other hand have sport as their priority and accept environmental and nature protection measures as long as they do not seriously (that is noticeably, but not painfully) curtail their activities.

When using this idealized sketch of the positions taken by environmentalists and sports enthusiasts, we must bear in mind that not every individual and institution acts in complete accordance with them. There are many advocates of environmental protection and nature conservation within the sports community and vice versa. Similarly, sports organizations and environmental groups accept various levels of compromise between these two poles.

In the search for compromise a great role is played by the values that each side, the sports and the environmentalist community, considers its **"core demands,"** which are not negotiable, and those it regards as less important, thus negotiable, demands.

For conservationists a core value or demand is that the habitats of rare plant and animal species (which appear in the Red List as in danger of extinction or threatened) be shielded from serious disturbance. For sports enthusiasts, however, it is important that they be permitted to engage in their activity in such an environmentally sensitive area, even when the time or location of the activity has restrictions placed on it.

When conflict over sports activities in environmentally sensitive areas arises, however, sports enthusiasts can move their activity to an area that is not threatened. Plant and animal communities needing protection cannot be moved, as they are tied to specific biotopes. Sports activities thus have flexibility with respect to their site and so can harmonize with the environmentalists' core demand for protection of sensitive sites. The exercise of the same sport within the larger surrounding area is thus in no way questioned or compromised.

Conditions for accepting limits: ground rules

Representatives of sports interests and those representing environmental concerns should always strive to deal with each other fairly when working out conflicts. It is legitimate to spar when working out positions. Only when scientific facts are confused with personal values do discussions become bitter. The door is then wide open to polemics. Therefore, the point of departure for all discussions must be to first determine the cause and effect relationships that have given rise to a given problem and then to debate how to organize and evaluate these findings: Where are the "core demands," where is there room for negotiation?

Sports enthusiasts and conservationists (as members of both non-government organizations and governmental agencies) agree on the fundamental point that sports activities must be limited in certain areas in order to uphold the recognized goals of nature conservation.[1] Disagreement nearly always arises over questions such as,

- "What kinds of restrictions are necessary to achieve
- which nature conservation goals
- for which area?"

In order to ensure the acceptance of necessary protection measures, the advocates of nature conservation must strive above all for credibility. Whether a particular environmental protection measure inspires confidence or not depends on whether the following rules are observed:

- All ecologically justified protection measures that entail a limitation of sports activities need to be discussed in a timely fashion with those whom they will affect.
- The use of an environmentally sensitive area by sports enthusiasts may not be judged in isolation from other uses of that area.
- Sufficient scientific data must be available and carefully evaluated so that a workable decision can be made.
- In evaluating protection measures those are to be preferred which both ensure protection and place the minimum of restraints on use of the area by sports enthusiasts.
- It is preferable that sports enthusiasts adhere to limitations on use out of a sense of personal responsibility rather than enforcing such limitations through governmental authorities, as long as the measures are not thereby compromised. Periodic and appropriate checks (monitoring), however, must take place to see if people are observing the measures.

---

1 In a survey of all sports federations that promote the different activities outlined in this handbook, representatives from them unanimously agreed that limits on sports activity are acceptable when these have a solid foundation (survey conducted for the updating of the manual in 1996).

5. Areas Tolerating Different Degrees of Environmental Stress from Sports Activities

5.1. Sensitive Areas Worthy of Protection

Whether a sport activity or sport facility is acceptable from an environmental point of view depends on whether it negatively affects the quality of the landscape to a significant degree. The degree of adverse effect depends in turn on two things: on the kind and intensity of the proposed use, and on how sensitive and worthy of protection the area is. Ecologically valuable land that has been classified as such, according to established criteria of nature conservation, merits such protection measures. Such areas have become relatively scarce in our landscape of intense cultivation, and without protection their continuing existence is threatened.

The German Minister of the Environment, Angela Merkel, had the following to say on the necessity for a consistent and clear policy of nature conservation: "The threat to biological diversity is worrisome. The loss of species is irreversible: man cannot recreate extinct species. Further efforts towards the protection of species are necessary – including in Germany – to preserve biological diversity (cited in "Natur und Landschaft," vol. 1, 1996). The setting aside of areas as reserves is an important instrument of habitat and species preservation. Biological diversity can, moreover, be furthered outside of protected areas, for example, by decreasing the use intensity of agricultural land.

The threats to species come above all from increasing destruction of the habitat of wild animals and plants, from its reduction into small, separated pockets and its consequent loss of value. Tourism (and in this connection, sport activities) contributes in great measure to the threat facing biotopes and species, as it often makes use of valuable and sensitive land (such as shorelines and riverbanks, small streams and lakes, coasts, rocks, and areas providing refuge for wildlife).

The following types of habitat are regarded as especially worthy of protection (detailed statement of criteria in Schuboth, 1996):

Freshwater, coastal and alpine biotopes meriting protection:

- Tidelands and mud flats, meadows with springs, coastal salt marshes, coastal dunes and tidemarks in a near-natural state, rocky coastlines and coastlines with cliffs
- Bodies of standing water (pools, ponds, lakes, reservoirs) including their shorelines and zones of sedimentation
- Non-canalized stretches of rivers and streams in natural condition, in particular their deltas, sloughs, and oxbows
- Springs and their surrounding meadows and bogs, limestone beds with their adjacent vegetation
- Alpine meadows, screes, alpine hollows with extended snow cover, forests of dwarf trees growing above the tree line.

Biotopes needing protection that occur in areas predominantly used by agriculture and forestry in plains and hilly areas (excluding the aquatic biotopes mentioned above):

- Raised bogs, mires and fens
- Short and tall sedge marshes, beds of tall reeds
- Humid and wet meadows as well as meadows and pasturelands with variable groundwater levels
- Natural inland salt deposits and inland dunes not covered by vegetation
- Meadows and pastures on nutrient-poor soils, dry grasslands, heaths composed of various shrublets and juniper
- Various plant communities comprising grasses, heaths, and shrubs found in rocky areas; shrub communities growing on slopes or in deposits of boulders; vegetation occurring on gravel deposits or scree slopes
- Hedges and old hedgewalls, coppices
- Woodlands in a close to natural state, particularly oak woodlands, dryland forests, mixed oak-hornbeam forests, woodlands with orchids, beechwoods, heath forests occurring in steppelands or snowy areas
- Seasonally flooded riparian forests growing along streams and rivers, swamp- and bottomland forests with the characteristic vegetation found along their edges and in clearings
- Old-growth forests, heaths, parks and cemeteries in urban areas with significant tree and shrub cover.

Since 1992 the FFH Guideline (Flora-Fauna-Habitat) of the European Union has been in effect, the first comprehensive law governing the protection of habitats and species (Ssymank, 1994). To fulfill the Guideline's goals, a coherent ecological network of special nature reserves is to be created bearing the name "NATURA 2000."

5.2. The Recreational Landscape under Stress: Classifying Landscapes and Formulating Planning Policy

A rough division of the landscape used for recreation into types, according to the land's sensitivity and need for protection, can serve as a guide where conflicts are to be expected between the desire for unregulated recreational use and the need for conservation.

The following **types of landscape** are defined according to their **ecological sensitivity and need for protection** with regard to the potential impacts brought on them by sport and recreational uses. These are "tabu areas," "natural recreation areas, "and "scenic landscapes" (for more detail see Schemel, 1987).

Tabu areas are particularly valuable areas from an ecological point of view, and ones that are easily disturbed. They require a high level of protection that can only be achieved when strong restrictions (such as declaring an area off-limits) are imposed, thus excluding all potentially disruptive activities.

When all types of recreational activity (with the exception of maintenance activities in the interests of preservation) pose a threat to a given area, it should be designated as "tabu." In strictly protected landscapes (such as national parks), areas should be set aside relatively often as tabu zones.

Natural recreation areas are near-natural, ecologically valuable areas where the protection of nature must have priority before any uses. In these areas, however, the needs of protection are compatible with certain quiet leisure activities[2] such

2 "Quiet recreation" here means activities that do not use motors or require infrastructure (with the exception of trails). That quiet recreation is not always compatible with nature will be discussed in detail later in this handbook. This form of recreation, however, offers unique opportunities for conflict avoidance.

as hiking, bicycle riding, canoeing, rock climbing, swimming, nature watching, cross-country skiing, and fishing. Natural recreation areas thus react less sensitively to recreational uses than do tabu areas. They consist of various zones admitting different degrees of outside stress. Measures can be put in place to channel visitors and so keep recreational activities away from more sensitive areas. No buildings of any kind are to be allowed in natural recreation areas. Natural recreation areas may most often find a place in ecologically rich and complex landscapes, in relatively undisturbed forests and in biosphere reserves. They may also be located in more strictly protected areas such as national parks and protected habitats. The proposed recreational uses to take place in these areas, however, may not compromise the goals of nature conservation.

Natural beauty and suitability for a variety of sport and recreational activities characterize **scenic landscapes.** They tolerate a higher level of outside stress. In such areas nature conservation is not a priority, but rather a general objective to be taken into consideration when engaging in sports and recreation activities.
Land marked by forms of human cultivation such as agriculture and forestry are characteristic of this type of landscape. To be regarded as scenic landscapes, however, they must also show a relatively complex structure and be suitable for recreational activities. In such landscapes recreational activities may be concentrated in a few areas or may take place over more extensive areas without any restrictive measures.

5.3. Assessing the Need for Protective Measures

A central issue in most conflicts between sport and the environment revolves around the question of how much the area in question needs protection. In general, people do not question the environmental and natural value of areas under strict protection (such as national parks). Nevertheless, the measures in effect there are not appropriate for all areas. This is because protection measures only become necessary when activities and their related effects would threaten the valuable qualities found in a given area. Environmental protection thus presupposes a given ecological sensitivity (a predisposition to damage or stress).

For example: an ecologically valuable area with crags and rock formations can serve as an arena for horseback riding because this activity has no effect on a nesting ground used by peregrine falcons located on a particular rock. This same area, however, must be off limits to rock climbers (at least during certain times of the year).

The following must be considered when assessing an area's need for protective measures.

In this process one must always keep scientific data (information on causes and effects) separate from the value assigned to them (how the data will be weighed with a view to reaching certain objectives). In evaluating and resolving specific conflicts, the following steps should be regularly followed:

- The area or areas that sports enthusiasts want to use should be evaluated according to their varying need for protective measures.
- Each proposed activity should be evaluated for its potential to cause environmental stress or damage.
- The ecological sensitivity of an area must be assessed in relation to the proposed activity, and then the specific protection measures for the area determined.

A word on the distinction between the concepts "worthy of protection" and "needing protection": With "worthy of protection" we mean that a particular good or quality (for example, a certain animal species in an aquatic habitat) has been recognized as ecologically valuable and endangered. For conservation experts a given – normally infrequently occurring – ecological good or quality is considered to merit protection when these two criteria, high value and danger of disappearance, are present. A judgment that an endangered species (for example, one included in the Red List), a natural phenomenon, or a type of land is worthy of protection is not made for every single tract of land. It is rather determined for the country as a whole, for example in the form of "biotopes worthy of protection".

The following scientific criteria are used to define and delimit a protected area: rarity of a species or habitat, uniqueness, closeness of an area to a natural state, biological diversity, typical characteristics represented by a habitat or its species composition, and the degree to which an ecosystem can be restored.

Using these criteria we can determine whether a given species or habitat is worthy of protection. This process, however, needs to be broadened. In a specific case of conflict both the area itself and the recreational activity that is proposed for it must be carefully considered in its component parts. Only then can the real need for protective measures be determined. Not every natural good or value that is worthy of protection requires protective restrictions against every kind of recreational activity. Such measures are only called for when the good or value worthy of protection would be harmed by a particular recreational activity.

The outline below illustrates the process of evaluating and planning recreational activities to take place within areas that are in a close to natural state and worthy of protection:

Taking Stock

Key questions to consider

- What natural benefits and qualities worth protecting are present (general conservation objectives and specific reasons for protection: species, biotic communities, land formations)?
- What sport activity takes place or is proposed (type of activity, amount of the total area to be used, times when the activity will take place, number and behavior of people involved)?
- What qualities worthy of protection will be affected by the activities (potential points of contact)?
- To which non-sport related outside influences are the qualities worth protecting exposed (for example, disturbances or risks stemming from agriculture and forestry, fishing, hunting and the like)?

Evaluating Conflicts

Key questions to consider

- Which natural benefits or qualities could be compromised by which activities (nature and extent of the potential negative effect, conditions and policies that would limit or increase this effect)?
- With respect to nature conservation, is the sports activity in and of itself problematic? Can conflict be avoided or at least minimized by modifying the sports activity or by altering conditions or policies?
- How great is the protection interest on the one hand and the sports interest on the other (core interests, room for negotiation, possibilities for compromise)?

Formulating Measures

Key questions to consider

- Which measures (or alternatives) could completely eliminate or minimize negative effects?
- Are already existing conservation regulations being applied? Where are they not properly implemented or followed?
- How can the continuing enforcement of necessary protective measures be guaranteed?
- Are (further) protective regulations that have the force of law necessary, or are voluntary agreements sufficient? How will the degree of compliance be measured? Can we expect cooperation or are problems of enforcement likely?
- In what way could sports organizations themselves oversee the introduction and monitoring of the measures? What methods and policies for ensuring compliance will be possible?

Monitoring

Key questions to consider

- Have the agreed-upon protection measures actually been implemented?
- Have the measures actually brought about the expected result?
- Do stricter measures need to be taken? Can certain restrictions be relaxed without thereby endangering the desired protection?

6. Putting Protective Measures into Practice

It is, of course, a long way from realizing the need for change to bringing it about, but the process normally takes the following course:

- **Knowing** about the environment ("I am aware," "I know about cause and effect relationships and about my own role in the destruction of the environment")
- **Experiencing** the environment and caring about it ("I am seeing environmental destruction and it bothers me")
- **Having values** oriented towards the environment ("In this case, I consider the protection of nature and the environment more important than. . . .")
- **Intending** to act in an environmentally beneficial way ("I intend to do something when I get a chance", "I am prepared to do without something for the good of the environment")
- **Acting** in a way beneficial to the environment ("I am doing something for the environment and for nature", or "I freely accept limits in order to protect nature").

These steps characterize an environmental consciousness that progressively grows and leads a person to become actively involved in protecting nature and the environment. People's general acceptance of the need for environmental protection often rests on principle and does not include the last step. This process of environmental consciousness should rest on a person's understanding and spring from his or her own initiative. It may at times be necessary, however, to force socially desirable behavior (through laws), or to encourage voluntary changes in behavior (by means of incentives).

Ways to influence the behavior of sports enthusiasts in the environment vary according to the degree that they are based on voluntary cooperation or on enforcement by outside authority. These are:

a) Prohibitions and regulations (regulatory law)
b) Appeals to self-regulation (by providing information and relying on the individual's understanding of the need for regulation)
c) Voluntary commitments to act responsibly
d) Incentives (policy changes).

These four ways of guiding sportspersons' behavior, which are explained in more detail below, assume that individuals have a certain level of **environmental consciousness**. An awareness of the value of nature and the environment and of the importance of protecting them stands as the decisive motivator of behavior.[3] Such awareness can of course be fostered "from the outside" through providing information, yet is most effective when it comes from the individual himself and so results in behavior motivated by the person's own system of values. The following discussion, however, deals with "external motivators," that is the influence of institutions on the sports enthusiast's behavior.

a) Prohibitions and regulations assume that ecologically desired behavior will not voluntarily come about to the degree that is necessary. Rather, such behavior must be forcibly coerced by means of sanctions (for example, fines) and by administrative policing (for example, requirements that people stay on marked paths, declaring areas off limits to cyclists or hikers).

b) Appeals to voluntarily regulate behavior assume that sports enthusiasts will abide by reasonable and sensible restrictions. Appeals to environmentally responsible behavior, however, remain ineffectual when only a few recreationists heed them and a large number ignore them.

Above we pointed out the gap between environmental consciousness and environmentally responsible behavior (by no means found only among sports enthusiasts). A study conducted by the German Federal Environmental Agency (Umweltbundesamt) showed that environmentally beneficial forms of behavior are most likely to come about when suitable venues for communication (such as discussion forums and round tables) are provided where the parties involved can voice their concerns and inform themselves. In these settings participants can learn and exchange views about the positive and negative effects of various kinds of behavior on the environment and overcome barriers to communication and learning. Communication and information at the local level can bring about serious thought on specific environmental problems and their causes. It also makes it less likely that people will deal with such problems in a simplistic way, such as by placing the blame on the opposing side (UBA 1996).

---

3 Theoretically, in the case of prohibitions and regulations, the behavior of recreationists could be modified without an environmental consciousness on their part. This, however, would entail the threat of legal penalties and a high level of policing.

c) A policy promoting **voluntary compliance with rules of conduct** is a sensible way to avoid strict regulations backed by the force of law and enforced by conservation officials. This policy also gets around the danger that park users will ignore appeals that lack legal consequences. In each case the governmental authority responsible for protecting the nature reserve, and the sports organization concerned, must agree upon such a policy (for example, by signing a legal contract). Sports associations can in this way perform important functions. In specific cases of conflict they can take on the role of speaking for sports enthusiasts who are not members of the group as well as for commercial sports operators. They may thus find themselves assuming the responsibility for seeing that all sports enthusiasts adhere to the agreed upon rules. The governmental authority responsible for the park or reserve must, however, continually check to see that the policy of voluntary compliance is working.

d) Incentives for desirable behavior are not based upon the sports enthusiast's sense of ethics or feeling of responsibility for the common good. Instead, they appeal to people's desire to achieve a certain goal with the least amount of effort (such as cost or inconvenience). Thus it is wise to create policies that will bring about desirable behaviors from sports enthusiasts. For example, one could offer public transportation to the site of the activity that is cheaper and more convenient to use than the automobile.

This policy of influencing behavior "via the wallet" can also play a large role in the area of sport-related resource consumption, for example, when it fosters the environmentally sensitive use of materials in the manufacture of sports clothing and gear. Whether technologies and methods for saving energy and water and for generating less waste can establish themselves in the market (instead of just surviving in small niches) depends above all on the question of cost.

II. SPECIAL SECTION

PROTECTING NATURAL RESOURCES IN SPORTS ACTIVITIES (A)

1. Environmental Significance

The exercise of sport can affect or use the following environmental media and natural resources: climate, air, soil, water, fauna and flora, raw materials.

Several **alternatives and strategies** exist in all types of sport and recreation that allow the conservation and more intelligent use of natural resources, for example,

- in the use of sports gear and clothing,
- in the construction and operation of sports facilities,
- by more intense use of existing sports facilities (for example, through consolidation of facilities),
- in holding sports events and competitions,
- in transportation to and from the site of a sports event.

Conservation and intelligent use of natural resources, while not an end in and of itself, serves the interest of sustainability

- by preventing an impending climatic catastrophe through the decrease of CO2 emissions,
- by protecting natural resources present in only limited quantities or by limiting their use (for example, ground water and fossil fuel sources),
- by decreasing negative effects on air, water, and soil, specifically damage caused by the production of energy (from coal, oil, gasoline) and by waste disposal (in landfills or incineration plants).

The **goal** of attaining **sustainable use of environmental resources** may be reached through:

- saving energy and using alternative (renewable) sources of energy,
  reducing the consumption of drinking water by using rainwater where possible (for example, for irrigating outdoor sports grounds),
- treating the soil with care, protecting it from erosion and shielding it from damaging additives (for example, by not using chemical fertilizers and pesticides, or reducing their use on outdoor sports sites) and avoiding the paving of large surfaces,
- avoiding sewage as far as possible and treating it in appropriate facilities (thereby protecting surface and ground water from contamination),
  generating the least amount of waste possible and recycling trash (for example, through reusable containers, or by using organic waste for compost which in turn could be used in the sport facility's green areas),
  using materials in construction and in the manufacture of sports gear and clothing that are not harmful to the environment (recyclable materials, avoiding hazardous waste),
- giving preference to non-motorized forms of transportation in order to decrease noise and pollution, and enacting policies that encourage their use by athletes and spectators.

Problems of land use posed by sports activities practiced in the landscape or in facilities can be solved by choosing an environmentally suitable site for the activity or facility. The problems we shall consider here, however, arise from the use of resources, from the "mundane" aspects of sports activities, ranging from clothing to heating of indoor spaces to the flushing of toilets. Recommendations that are valid for individual households and for general everyday behavior will be looked at here with respect to the situation of the sports enthusiast. The discussion will consider sports organizations in particular, for they have numerous opportunities to put these recommendations into effect, ranging from the mutual exchange of information on environmentally beneficial policies and strategies to exemplary behavior that catches public attention. The fact that many such policies benefit the wallet as well as the environment can only encourage environmentally sound behavior.

2. Environmental Management in the Sports Organization

The sports organization offers an excellent arena for members to learn about issues of nature conservation and environmental protection. This in turn can influence sports enthusiasts to act in a more environmentally considerate way. Moreover, by offering a good example of environmentally responsible behavior, the organization can motivate sports enthusiasts to consume fewer resources both when practicing their sport and when engaging in everyday activities. What exactly can the organization do to bring about these effects?

Below we outline some steps toward an environmental management system that seeks to expand the responsibilities of sports organizations in order to bring about a steady improvement in environmental protection.

- **Formulating Environmental Guidelines**
  The organization can in its charter or constitution commit itself to the principles of a sustainable use of natural resources and responsible land use. In addition, the organization can set specific guidelines for itself (which go beyond the principles spelled out in its charter) as part of an "environmental policy." It thereby sets binding goals for itself in each of its activities or departments.

- **Carrying Out an Environmental Check**
  An environmental "test" should ascertain where the organization could act to the benefit of the environment and where it could make improvements to achieve its environmental policy goals.

- **Putting Together an Environmental Program**
  In this step the organization formulates measures which it feels are reasonable and feasible, that seek to make up the deficits that were uncovered in the environmental check. The measures should also take advantage of opportunities for policy-making indicated by the test. Measures may be organized according to priority (urgent or less urgent need for action), a time schedule (desired in the short, middle, or long-term) and their associated expense (reduces or increases costs, has no effect on costs).

- **Implementing the Environmental Program**
  The sports organization must supply the required organizational structures, personnel, and money necessary to ensure that the proposed measures do not just remain on paper, but are steadily put into practice. The organization should appoint a **person in charge of environmental matters** who feels personally responsible for implementing the environmental program and its further development. This person should have a place on the executive board of the organization. He or she will need support "from below," for example from an environmental panel or working group. Regular information (provided, for example, in the organization's newsletter or "bulletin board") on the necessary steps the organization is committed to take, on what it has already accomplished and on what still remains to be done, can further the implementation of the environmental plan. Periodically the organization should run a performance check (interim assessment) to see how the plan is working. The organization should always consider if further measures would be sensible and feasible and revise the environmental program accordingly.

- **Public Relations**
  People both in and outside of the organization should know about its work on behalf of nature and the environment. This increases the public's esteem for the organization and impresses, and perhaps influences, sports enthusiasts who do not belong to the organization. It also may cause members of the organization (perhaps the more idealistic among them) to identify positively with their organization. The organization also places itself in the public eye by taking part in political decisions made at the local level that either directly or indirectly affect sport and the environment.

3. Checklist for Natural Resource Protection (Energy, Water, Waste)

Here we discuss in outline ways to sensibly use natural resources that will not be covered in the chapters entitled "Large Sporting Events," "Sports Fields and Facilities," and "Gymnasiums." Our object here are facilities operated by sports clubs.

- **Saving Energy**

  What possibilities do we have here for saving energy? First the total energy consumption of the facility must be added up and then possibilities for savings (in electricity usage and heating) determined. One has to consider, for example, the heating of changing rooms, floodlights, pumps (for irrigating playing fields), the apartment for the groundsman (heating, insulation), restaurants and cafeterias with their associated equipment (lighting, refrigerators, dishwasher), and miscellaneous sources of consumption (such as washing machines and clothes dryers).

  Questions for Finding Ways to Save Energy:
  - Can fossil fuels be replaced by renewable energy sources (solar energy, heat recovery, doing without indoor electric heating)?
  - Can better insulation be used in remodeling existing buildings or in new construction?
  - Are rooms being overheated or improperly ventilated?
  - Can surplus heat (for example, from shower water or cooling systems) be used?
  - Can power surges (which result when too many appliances are in use at the same time) be avoided?
  - Can lighting be lowered or better utilized?
  - Are environmentally friendly forms of transportation being used as much as possible?

- **Conserving Drinking Water**

  Sources of usage (identify current usage and where it originates):
  - Locker rooms: from showers, sinks, flushing of toilets
  - Meeting rooms: washing and toilet facilities (see above), kitchen operations (cooking and washing dishes)

- Sports-related washing and laundering: washing and toilet facilities (see above), washing gym clothing and cleaning athletic shoes
- Sports facilities (upkeep): irrigation of outdoor playing fields and green spaces

Questions for Finding Ways to Save Water:

- How can the consumption of drinking water be reduced or replaced by rainwater?
- How can we change our individual habits so as not to waste water (such as shutting off faucets immediately after drawing water)?
- Where is better maintenance called for (for example, fixing leaking faucets and showers)?
- Are available water-saving devices being utilized (for example, toilet handles allowing short flushes, low-flow showerheads)?
- Is rainwater being used (rainwater collection for irrigation of green spaces)?

- **Disposing of Waste in an Environmentally Conscious Way**

  How can we avoid generating waste altogether and reduce the amount of trash which cannot be recycled? First we have to find out where waste is generated (identifying existing sources, finding possible solutions).
  The most important goals are:
  - avoiding unnecessary and wasteful use of materials (leads back to problem of "saving energy"),
  - recycling trash: finding ways to introduce garbage as a "raw material" back into the economy or into natural cycles,
  - avoiding hazardous waste (by simply not using products containing such substances): its disposal entails great risk to the environment (because of the as yet unclear consequences of burning or burying it).

The above steps relieve stress on the environment because they mean less need for landfills, fewer incineration plants, less pollution of air, water, and soil with toxic substances, less burning of fossil fuels, and decreased emissions of CO2 (threat to global climate).

Some practical hints:
Priority should be given to cutting down on packaging and to selecting products with a longer life span. In this way the volume of trash can be reduced.

- Reusable containers are to be favored over single-use containers (such as plastic bottles, beverage cans and packaging made of composite materials).
- Reusable dishes should be used (at sporting event food concessions) instead of single-use disposable ones.
- Products with excessive, unnecessary packaging should be rejected.
- Purchasing decisions should take into account both the type and quantity of packaging that was involved in the transport and delivery of the goods.
- Individual serving-size packaging should be replaced with large containers (metal bowls or canisters) that can be reused.
- Products with a longer life span should be selected.

Waste that is unavoidable can be disposed of in the following ways:
In accordance with local arrangements and provisions for garbage disposal, each type of trash should be thrown away into a container designated just for it. In this way we can ensure that it is recycled as far as possible. Roughly speaking, trash consists of the following types: valuable materials that can be recycled and used in the manufacture of new products (for example, paper, glass, various plastics, metals), organic waste that can be turned into compost and returned to nature (vegetation of all kinds, unused parts of fruits, vegetables and other edibles), trash that cannot be avoided or recycled, and finally, hazardous waste, whose proper disposal presents a host of difficulties (often burned in various kinds of incineration plants with consequent risks to the environment), such as spray cans, unused paints and varnishes, oil-soaked rags, incompletely emptied paint containers, oil filters, batteries.

Clearly identifiable trashcans should be placed wherever garbage is commonly generated. The cans will then be emptied into larger collection containers.

It is important that club members feel motivated to separate their trash. One way to bring this about is to provide information each year on the amount of non-recyclable waste generated, so that members can see the club's successes in avoiding trash and recycling it.

4. Environmentally Friendly Transportation

Mobility has notably increased in the last decades in the field of recreational activities as in all areas of life. Traffic stemming from leisure activities, whether on day trips or during vacation, made up 54% of the total traffic generated in Germany in 1994, more than that originating from commuting and shopping combined. Mobility becomes a problem when it brings stresses on people and the environment.

In the field of sport and recreation mobility arises when the sports enthusiast travels to a sports club, facility, site, or a sporting event and then from there returns home. This process concerns not only the transport of the sports enthusiast himself, but of his gear and equipment as well.

What possibilities does the sports enthusiast have when he wants to practice his sport without causing substantial stress to the environment? The most important decision concerns the choice of transportation: namely, using **environmentally friendly forms of transportation** (train, bus, bicycle), or walking, when this is possible. He can also decide to pick a place close to home to exercise his sport, and thus avoid long distances. This **"close-to-home"** type of sport activity of course depends on there being a variety of sports sites or facilities in the person's neighborhood.

Appeals to sports enthusiasts to travel in a more environmentally friendly way are not enough, however. It must be made easy for them to choose public transportation or the bicycle over the car. The responsibility for creating and maintaining the necessary conditions for this lies with urban and transportation planning agencies. They must see that

- sports facilities are built close to where people live,
- sites for sport and recreation can be easily and inexpensively reached by bicycle, bus and train.

Sports clubs and the operators of sports facilities and recreational attractions can contribute to the creation of these necessary conditions by, for example, simultaneously offering secure and convenient bicycle racks and reducing the number of parking spaces. A sports club can also organize the transport of sports gear and equipment and offer storage space in its facilities. Public transport authorities should allow sports enthusiasts to take along their gear.

The best way sports enthusiasts can reduce transportation-generated stress on the environment is to act in a responsible way whenever conditions permit, that is, to avoid unnecessary car trips. Appeals, information and setting a good example can all play an important role here. Sports enthusiasts can also plan to car-pool with others wanting to drive to the same site.

SOUND AND NOISE GENERATED BY SPORTS ACTIVITIES (B)

1. Environmental Significance

Numerous sports activities involve sounds that others can experience as irritating noise. This is particularly the case when areas requiring a good deal of peace and quiet, such as residential areas, border on the site of the sports activity. This desire for tranquility is most acutely felt in the evenings and at weekends and holidays. Noise is also a significant problem during large sporting events (see chapter C).

Sport-related sounds originate from

- sports equipment (such as motors and balls),
- sounds made by players themselves (yelling),
- sounds made by technical media equipment (loudspeakers, starting pistols),
- sounds coming from spectators (applause, booing and catcalls),
- arrival and departure of players and spectators (traffic sounds).

2. Terms and Measurement Procedures

Sound is a phenomenon of vibration (in this case through the air) that travels in the form of acoustic waves and that can be measured as acoustic pressure at the point of reception. Sound intensity is reckoned in decibels (dB) using a logarithmic scale. When this intensity is related to human hearing (calculated according to an internationally recognized standard frequency value curve, "A") we have the perceived sound intensity in dB(A).

Noise is defined as sound that disturbs (endangers, negatively affects or considerably stresses) neighbors or third parties.

Since sound intensity is logarithmically calculated, one has to be careful when evaluating increasing or decreasing levels of noise. The following rules should be kept in mind:

- Two equally strong sources of sound added together (thus a doubling) equal an increase in intensity of 3 decibels; halving this amount constitutes a decrease in 3 decibels. Example: The confluence of two sources of noise of 50 dB each equals not 100, but 53 dB.
- When two sounds of unequal intensity converge, a value between 0 and <3 decibels (according to one formula) is added to the greater of the intensities. A sound more than 10 dB less in intensity than another contributes only minimally to the cumulative intensity of the sounds and can thus be ignored.

The following rule of thumb holds true when measuring the effect of noise on people: An increase in sound intensity of 10 dB(A) is perceived as a doubling of volume. This is true only for intensities of momentary duration (and so not for averaged intensities extending over a period of time).

When sound is allowed to travel unobstructed from a fixed point its intensity decreases by 6dB(A) for each doubling of the distance between the source and the hearer. Barriers (walls, earthen berms, vegetation) can, however, have a more or less strong influence on the outward dispersion of sound.

The following table gives the intensity of sounds resulting from various forms of human communication [emissions values in dB(A)] measured 100 m from a sports facility (source: BIS 1994).

Sound	dB(A)
Loudest possible shouting Football coach shouting	62
Average shouting per person when goal is scored	55
Shouts Spectators cheering at players Players cheering at other players Trainer on a tennis court Calls between players	38
Normal communication Players on a tennis court Expression of disappointment at football game by sighing aloud	33
Clapping: - very loud - normal	43 38
Umpire whistling	67
Children shouting (in school yard)	30

The sharpness of an acoustic impulse (for example, of tennis balls) and the information that recreation-related sounds convey to the hearer make noise even more annoying. To compare different types of noise with each other (or with standard emissions levels or values used in planning) one uses the **evaluation intensity** measured at the place where the sound is heard. This measurement is composed of the **measured value and adjustments.** Adjustments (in the form of additions to the intensity value) may be necessary to take into account the sharpness of an acoustical impulse or the message a recreation-related sound conveys to its hearers. They may also be made when sport activities occur during quiet daytime hours or at night.

Standard Impact Levels as Set by the German Federal Noise Control Regulation for Sports Facilities (Sportanlagenlärmschutzverordnung)

Area	Time of day	Maximum permitted noise level in dB(A)
Industrial and commercial areas	daytime, outside of quiet hours daytime, during quiet hours night time	65 60 50
City or town areas, village areas, mixed areas	daytime, outside of quiet hours daytime, during quiet hours night time	60 55 45
General residential areas, small housing estates	daytime, outside of quiet hours daytime, during quiet hours night time	55 50 40
Exclusively residential areas	daytime, outside of quiet hours daytime, during quiet hours night time	50 45 35
Health resort areas, hospitals and other care facilities	daytime, outside of quiet hours daytime, during quiet hours night time	45 45 35

3. Ways to Avoid or Decrease Noise

3.1. The "Subjective Factor"

The extent to which noise bothers a person depends not only on the physically measurable intensity of the sound, but also to a high degree on the situation and attitude of the noise sufferer.

Before we discuss measures that combat noise by means of planning, technology, and regulations, we want to emphasize the importance of good stress-free relations between a sports club and the neighborhood where it is located. Besides being desirable, a neighborly and friendly atmosphere acts to reduce the sensitivity to noise that the neighbors of a sports facility experience.

The following **factors of subjective noise perception** come into play when people feel subjected to sounds and noises against their will:

a) the individual's state of being (sensitivity to noise, state of health, level of stress in his life),

b) the individual's relationship to the source of the noise (feelings of sympathy or antipathy towards it),

c) the activity the individual wants to engage in when the noise occurs (sleeping, relaxing, talking, listening to music, performing work requiring concentration),

d) whether the sound is necessary or not (a noise perceived as necessary and reasonable is less disturbing than a noise regarded as unnecessary and pointless).

Factors (b) and (d) make it especially clear that sports clubs have certain opportunities to reduce the sensitivity to noise experienced by citizens affected by their operations. Clubs can help to bring this about when they

- create an overall relaxed and amicable "climate" in their neighborhood (for example, by inviting nearby residents to events),
- emphasize the positive (health-promoting) function of sport and recreation,
- avoid making unnecessary noises (not giving the impression to neighbors that the club does not care if they are annoyed),
- take advantage of ways to reduce noise emissions (so that neighbors can see that the club respects their need for peace and quiet).

In its communication with its neighbors, the sports club should try to create a positive attitude toward the users of the facility, the facility itself and the sports that are played there. The sports facility should strive not to appear as an outside invader in the eyes of its neighbors.

3.2. Planning Measures

In cases where poor planning has led to serious noise problems, attempts to solve conflict have often proven insufficient as well as financially costly, and have entailed limitations on the club's operations. Thus the design and construction of new sports facilities present planners with important opportunities for preventing subsequent noise. Their designs should obviate as far as possible the need for later measures to combat noise.

The following measures can be useful, provided that an actual need for the facility exists (necessity for the facility as it is planned, having the right dimensions, fulfilling actual needs):

Site Selection

Avoid a site that would place noise-producing activities (such as certain forms of sport) next to areas requiring peace and quiet (housing). If such a location is unavoidable, one must investigate early on whether the planned facility would fit into the social milieu of the area or not (level of acceptance).

The amount of noise that a residential area already receives (for example, from traffic) also plays a role here. It can happen that noise emanating from a sports facility in an area with an already relatively high noise level would have little disruptive effect. A markedly louder noise occurring at the same time in such a situation has a "masking effect." Yet one must bear in mind that the traffic noise might die down in the evening or on weekends and so the masking effect disappear.

In addition, attention must be paid to

- buffer zones,
- placement of activities within the facility,
- making sure that plans are backed by laws
- "traffic-calming" (measures to reduce and slow down automobile traffic)
- involving sports representatives in the planning process.

3.3. Organizing Events and Activities

Here in an outline format we list ways of organizing sporting events and activities that should keep noise down. These measures restrict sporting activities to certain times and places as well as regulate adjacent land uses and certain conditions relevant to sport:

- Making public transportation more attractive, for example, by running trains and busses more frequently, and by allowing passengers to take sports equipment with them at low cost (avoiding automobile use)
- Scheduling (certain sporting events or activities are scheduled for times when nearby residents require less quiet)
- Limiting use periods (forgoing sports activities during certain quiet times)
- Limiting use of space (for example, holding training sessions on the side of a field furthest from nearby homes)
- Taking down football goalposts (dismantling or moving goalposts on public playing fields when noise becomes too great during times when quiet is required)
- Oversight of events by facility personnel and prohibition of alcohol (to prevent inconsiderate noisy behavior on the part of a few players or spectators).

3.4. Technical Measures

Measures to decrease noise can focus on the source (such as machines and equipment) or the path by which the noise travels (barrier walls and the like).

Measures applicable to equipment:

- Lowering the volume of loudspeakers (when making announcements at track and field events and football games)
- Decentralizing the placement of loudspeakers (putting them only in the areas for which the information is intended, thus allowing their volume to be set lower)
- Quieter starting signals (silencers on starting pistols)
- Fences that quietly absorb hits from balls (no clanging noises: use a fence made from synthetic material or place buffering material on posts of chain link fencing)
- Selecting quieter maintenance machinery (for example, machines for marking lines and boundaries on fields, lawn mowers)
- Sport equipment improvements (such as quieter motors for aerial sports).

Measures applicable to pathways of sound diffusion (vertical barriers):
Protective plantings: Although this land-intensive means of reducing noise often cannot be realized due to lack of space (especially in urban areas), plantings are particularly desirable. Besides reducing noise (although not by much) they fulfill other important functions (they reduce air pollution, protect from wind, act as a screen and provide a habitat for animals). Small green strips scarcely reduce noise, yet they put the source of the noise out of sight and have a possibly calming effect on neighbors.

Sound berms: A berm planted appropriately to its site is the best of all "technical improvements," since it fulfills both aesthetic and ecological functions (as a landscape element and as habitat). The need for a relatively large amount of land can stand in the way of its construction (it requires sides that slope at a ratio of 1:1.5 to 1:2 and that smoothly level out at the base). The land requirement is lessened when garages, storage sheds, changing rooms, etc. are built into the earth mound. With sound berms a reduction in sound intensity of 6-12 dB(A) (depending upon their height) can be achieved.

Berms can also be integrated into a system of foot and bicycle paths and recreational zones. The following combinations have proven effective:

- A berm in combination with situating the facility within a depression; when possible utilizing the natural slope of the land (berm can be built from excavated earth).
- Using the berm as a grandstand for spectators on the side facing the playing field (the crest of the berm lies above the level of the spectators).
- An earthen berm with a wooden wall on top.

When very little space is available, vertical walls with plantings or sound walls can be used instead of berms. When possible the sound wall should be covered with vines or at least have plantings on the side facing neighbors. A fence for catching balls has also been used to provide noise reduction (by using sound-absorbing mats) in combination with other measures. The closer the sound wall lies to the source of noise the more effective it is. Of course, its effectiveness is also determined by its height. The average decrease in sound intensity from sound walls lies between 6 and 12 dB(A).

LARGE SPORTING EVENTS (C)

1. Environmental Significance and Potential Conflicts

The greatest areas of potential conflict arising in large sporting events are at the same time the best places to start eliminating the negative environmental effects of these events. These are:

- Traffic-related problems: How can the use of automobiles and motorcycles (from people driving to and from the event) be kept to the lowest possible level?
- Waste-related problems: How can the food concessions be organized to generate the least amount of non-recyclable garbage possible?
- Site selection: From an ecological point of view, which location is the most suitable for the event? Where are new buildings and facilities unnecessary?
- How can we avoid damage to the landscape? Does public transportation serve the site?

Traffic noise and resource consumption (energy, water, generation of waste) constitute the main problems posed by sporting events in urban areas. Sporting events in the countryside, however, produce conflicts with various elements of the landscape, in addition to generating traffic noise. The following forms of environmental stress can be caused by setting up and operating temporary buildings, tents and grandstands, as well as by traffic to and from the event:

- Stress on the soil:
  - soil compression (from construction, vehicle tires, walking on sensitive subsoils),
  - entry of pollutants into the ground (trash, chemicals, oil and gasoline from vehicles, aircraft and machinery),
  - entry of organic wastes into soil (from trash, feces, urine), especially problematic in nutrient-poor soils

- Air pollution:
  - exhaust from vehicles and machinery

- Stress on flora and fauna:
  - damaging or destroying vegetation (either directly through physical impact, or indirectly through entry of nutrients into soil) by trampling, plucking or uprooting plants or parts of plants; loosening slopes; trampling vegetation on the edges of forests; damaging riparian plant communities; displacing species dependent on nutrient-poor conditions
  - causing harm to animal communities (either directly by frightening them away, or indirectly through permanent destruction of their habitat) by disrupting their breeding grounds, intruding into the safety sphere animals need around them, changing the structure of biotopes (for example, by clearing wooded areas or draining land)

- Dangers to human health and stresses on human well-being from:
  - harmful substances (for example, air pollutants)
  - noise (annoying to residents particularly during times of rest)
  - offensive odors (resulting, for example, from insufficient sewage treatment or waste disposal, smoke from barbecue grills, exhaust fumes)
  - disrupting the beauty of the landscape (from construction sites, parked cars, destroying vegetation in parklands).

2. Finding Solutions

a) Choosing a form of transportation

Here the goal is to motivate as many people as possible to use public transportation when travelling to a sporting event so that the stress caused by automobile traffic can be minimized. To achieve this it is above all necessary to make public transportation attractive in price and in its level of comfort and convenience so that it can compete seriously with the private car.

To increase the competitiveness of public transport and encourage walking and the use of bicycles, measures should aim at making the automobile more expensive and inconvenient to use. At the same time they should promote taking public transportation. Ways to achieve these goals include:

- Selecting a site for the event that is easily reached with public transportation. When this is not possible, special bus routes to the site should be set up.
- Setting up special arrangements with public transit authorities for transporting participants to and from the event (offering free or reduced local fares, special busses and trains, special bus routes; scheduling more frequent train or bus service, which reduces waiting times and increases both transportation's capacity and convenience; offering "combination tickets" to events, which include public transport fares).
- Allowing people to bring bicycles with them on public transportation.
- Providing special trains and busses (with low prices for participants, for example, to major league football matches).
- Linking public transportation with taxis and small busses, particularly at night.
- Setting up guarded parking lots on the edges of large cities or in places in the countryside that can support such a use (participants then travel by shuttle bus to the event or to public transport stops/stations).
- Passing out meal or snack coupons or raffle tickets upon presenting a stamped transport ticket.
- Encouraging people to walk or ride bicycles and making it easier for them to do so (setting up guarded bicycle racks close to the event; offering bike rentals and repair services; providing direct, secure, and well-marked paths for pedestrians and cyclists).

- Using non-motorized forms of transportation or vehicles powered by an environmentally-sound source of energy in the area around the event. Such transportation can also shuttle event participants to and from public transport stations, or to and from distant parking lots.

Ways to discourage travel by automobile:

- Offering only a few parking spaces and locating these far from the event site. Information on scarcity of parking should be made known to the public (or to special target groups) well in advance of the event (one exception: sufficient parking spaces for the handicapped should be provided).
- Employing security guards to prevent inappropriate parking of cars on sidewalks, in private roads and driveways, in green spaces and on farm fields. Roadsides and shoulders should be cordoned off.
- Charging high parking rates (income from fees can support the operation of shuttle busses).
- Closing off roads around the event site to private automobile traffic (residents of the area, however, receive passes).
- Closing roads leading to the event site when this is located in the open countryside (central parking lots should be located far from the event site and high rates charged).
- Engaging in public relations work that explains the reasons for discouraging commuting by car and shows transportation alternatives.

b) Dealing with Trash

Here the chief goal is to avoid generating garbage. Where this is not possible we should facilitate the recycling of trash, at least as a supplementary measure (using recyclable materials, throwing trash into separate containers). Finally, non-recyclable garbage should be completely disposed of.

The purpose of trying to avoid unnecessary garbage is not just to spare the event organizers trouble, but also to help solve the general "trash crisis" that currently exists.

Ways to design food concessions to generate less trash:

- Offering food and drink that require the least amount of packaging, dishes and cutlery.

- Using a system where participants place a deposit on dishes and silverware, which they get back when the dishes are returned; also selling dishes (bearing the event's emblem or symbol) as souvenirs.
- Prohibiting or restricting single-use containers for food and beverages, offering instead washable dishes and silverware as well as reusable containers for beverages, for example, returnable bottles with straws, glasses, juice and beer on tap served in glasses.
- Selecting foods and beverages whose preparation and storage produces the least amount of trash possible.
- Not distributing throw-away promotional articles (plastic bags, brochures, event programs and the like).
- Asking suppliers of foodstuffs to deliver them with a minimum of packaging (examples: wooden crates for fruits and vegetables, baskets for baked goods, plastic tubs for meat and sausage, jute sacks for potatoes, reusable metal cans for milk).

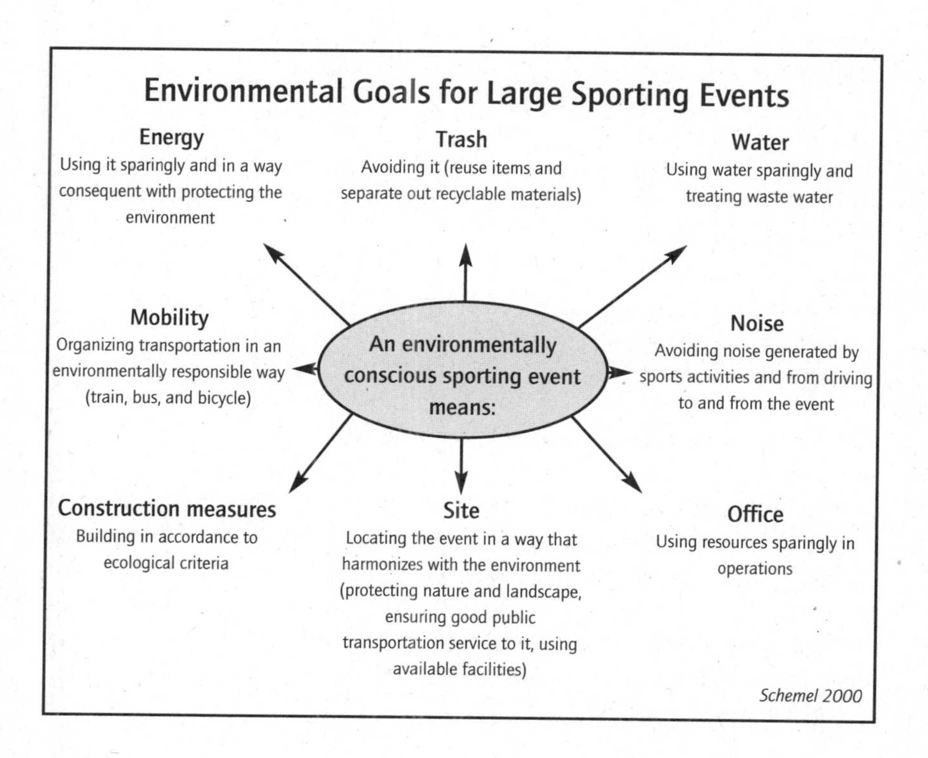

Ways to dispose of trash and sewage in an environmentally-conscious manner:
- Collecting all garbage (setting out sufficient trash containers).
- Collecting garbage according to its type (glass, paper, compost, hazardous waste, non-recyclable trash in separate clearly marked containers.
- Establishing a site for collecting hazardous wastes (for example, for oil and left-over gasoline in the case of auto races, lighting tubes, left-over decorating materials such as paint and glue, batteries).
- Installing devices that will separate grease and starch out of kitchen waste water.
- Reducing the use of chemical agents in the disposal of mobile toilet facilities. Compost toilets merit checking into (they have been successfully used in campgrounds).
- Connecting facilities to the local sewer system when possible.

c) Selecting a Site and Channeling the Flow of Visitors

Tips were given above on how to reduce the amount of auto traffic when selecting a site for a sporting event (placement near public transportation stops and stations).

In principle, sites that are home to sensitive plant and animal communities, or to individual species needing protection, are ecologically unsuited for sporting events. So are sites lying near such habitats that would experience stress
- from the sports competitions,
- from participants and spectators driving to and from the event,
- from activities connected with the event (sales stands and food concessions, pedestrians, dogs running around),
- from sounds and lighting.

Designated nature reserves are not the only areas that are sensitive and deserving of protection. Other areas with special ecological qualities are worthy of it as well, such as near-natural riparian habitats, forest edges, wetlands, bottomlands, and other habitats comprised in the "Mapping of Valuable Biotopes". Some habitats are only susceptible to disturbance during certain times or seasons (when birds are nesting, rearing hatchlings, or molting). The conservation authority in charge of the area in question should always receive timely notice of planned events, for they can determine if the site proposed for an outdoor event is ecologically suitable.

SPORTS FIELDS AND FACILITIES (D)

1. Significance for the Environment

The term "sports fields and facilities" encompasses large and small playing fields, facilities for track and field athletics, all types of facilities for recreational games and sports, as well as supplementary buildings and side areas.
Sports fields and facilities **affect the environmental quality** of areas surrounding them in the following ways:

- Which specific areas will be affected by the siting of a facility (will the ecological quality of these areas be harmed or enhanced)?
- Would adjacent residential areas be bothered by noises or lighting?
- Would pollutants be emitted?
- Will the paving of surfaces be held to the absolute minimum?
- Will consideration be given to maintaining unused or infrequently used green spaces in the most natural way possible?

2. Potential Conflicts and Ways to Solve Them

A sport field or facility is fundamentally a built-up space separated from nature. Ecological quality can only be maintained or developed to a very minimal degree in intensely used outdoor spaces, including mowed fields. (Unused side areas left in a close to natural state are not included here.) Therefore the siting of playing fields and sports facilities becomes extremely important when trying to avoid the loss of precious ecological qualities.

Conflicts can also arise when the facility causes disturbance to its surrounding areas, for example, when sports activities produce noise or when pollutants enter ground- or surface water. Sports activities must thus be organized, and facilities designed, to have a minimal disruptive effect on the surrounding environment.

Another cluster of problems arises from the use of resources (energy and materials) and the consequent air pollution and generation of waste (see chapter A). The use of drinking water for irrigating playing fields must also be regarded as a conflict.

Conflicts between sports facilities on the one hand and the needs of nature and the environment on the other can arise:

a) from the selection of a site that has particularly valuable environmental qualities meriting protection:

- destroying habitats that are in a near-natural state (for example, forest, fallow grassland, riparian areas, meadows, urban recreation areas in a relatively natural state),
- altering the natural topography of an area through excavating and depositing earth (a conflict occurs when the natural topography plays an important role for the ecosystem of the surrounding area, or when it contributes greatly to the visual appearance of the landscape),
- disturbing nearby residents with noise from sports activities (resulting from building the facility too close to existing residential areas),

b) from design, construction processes, and choice of materials that place stress on the environment:

- when a high percentage of outdoor space is bare of vegetation (lack of near-natural "side areas"),
- when paving covers a large portion of the surface area (as paved surfaces have low permeability, the replenishment of groundwater is reduced),
- from the use of materials containing potentially harmful additives (risk of water contamination from materials that leach harmful substances into ground- or surface water) or requiring a great deal of energy in manufacture and upkeep,

c) from a high intensity of grounds upkeep:

- selecting seed and setting a mowing frequency that result in playing fields lacking a variety of grass and turf species,
- from frequent applications of fertilizers, which then seep into groundwater or run off into surface water (since vegetation does not completely absorb fertilizer),
- using pesticides, which in addition to pests also affect other ("harmless") organisms and which can contaminate ground- and surface water,
- using drinking water for irrigation (instead of using rainwater or recycled water).

For the good of the environment pesticides should be used as sparingly as possible. A turf resistant to wear and to disease can be obtained through appropriate means (for example, fertilizers, more resistant grasses, and less intensive maintenance).

A site for a playing field should not be selected according to whatever the market makes available. Rather, a site should be selected within the scope of regional planning, wherein several alternatives for a prospective piece of land should be considered. Where possible an already stressed site (such as an industrial zone no longer in use) should be selected in order to preserve already scarce ecological qualities in urban areas. As land of low ecological value is often well suited for playing fields, local planning authorities should act in a timely manner to secure such land for this use. When possible, consultation between sports and environmental representatives should precede the acquisition of a site.

In its attempt to protect the quality of its groundwater, the government of the state of Baden-Württemberg has developed a guide for the adequate fertilizing of lawns with nitrogen. It gives the following amounts of nitrogen for mowed playing fields (providing sufficient fertilization, while minimizing risk):

Playing Field (arranged according to level of usage)	recommended amount of nitrogen per year
low level of use, cut grass partially removed	4 g N/m2
medium level of use, cut grass partially removed	12 g N/m2
high level of use, cut grass completely removed	24 g N/m2

Source: Ministerium für Ländlichen Raum, Ernährung, Landwirtschaft und Forsten, Baden-Württemberg, 1989 (Baden-Württemberg Ministry for Countryside, Food, Agriculture and Forestry)

Problems stemming from the **use of water** for playing fields and adjacent open spaces also admit of environmentally sound solutions. Here one should start by refitting irrigation systems to use rainwater or recycled water (which has been used for washing or other purposes) instead of drinking water. Drainage systems should be designed to collect water below the surface and put it to later use. Constructing and using an open, shallow well (groundwater of lower quality lying near the surface) also safeguards supplies of drinking water and as a wet biotope can fulfill ecological functions.

Unused side areas offer great opportunities for designing an environmentally-friendly sports facility. These spaces should be made as large as possible and allowed to develop in a manner that most closely approximates nature. The ecological value of a sports facility, and its ability to blend in harmoniously with its surroundings, depends to a large degree on these open spaces bordering playing fields. A diverse flora and fauna can develop much more easily in these spaces than in the small green strips one usually finds. Such open spaces can also provide other dividends, such as keeping the facility out of view and shielding neighbors from noise. To achieve this, side areas must be of sufficient size and protective berms should have the necessary height without being too steep. The maintenance of such open spaces should follow ecological criteria. Use of fertilizer and pesticides must be avoided altogether, and mowing should aim at an ecological optimum. A mosaic of spaces, each of which is mowed at a different interval, creates a variety of habitat niches for small animals.

GYMNASIUMS, INDOOR SWIMMING POOLS (E)

1. Significance for the Environment

Gymnasiums and indoor swimming pools touch upon environmental issues in the following areas:

- When selecting a site for these facilities it is important to find locations that do not negatively affect valuable recreational space and biotopes in urban areas. Sites should be easily reachable by public transportation or bicycle.
- In the construction and operation of the facility, specifically in the selection of building materials, in the construction process itself, and in the daily use of the building, attention should always be paid to the environment and human health (see chapter A, "Protecting Natural Resources").

Gyms and large indoor pools are used in the following sports:

Gymnasiums (not including indoor tennis courts or indoor equestrian courses)		
badminton	football	tumbling
basketball	gymnastics	trampolining
bicycle ball	handball	trick cycling
boxing	hockey	volleyball
budo	netball	weight lifting
fencing	table tennis	wrestling

Indoor Swimming Pools		
high-diving	swimming	underwater swimming

2. Potential Conflicts

The points outlined in this and in the following chapter ("Finding Solutions") are valid for other types of indoor sports facilities (when not otherwise noted), for example, for roller skating rinks, squash courts, ice skating rinks, and multi-purpose buildings. Gymnasiums and indoor sports centers (as well as other buildings) can have a variety of negative effects on the environment. These are listed below, organized according to site selection, construction process, and operations.

Selecting a Site

- Loss of green space having recreational and ecological value
- Devaluing neighboring areas in a near-natural state (through interfering with the soil structure, water regime, and vegetation)
- Dividing contiguous green spaces
- Causing negative changes in local climate (by acting as a barrier when positioned so as to block air movement through open channels, so-called fresh air cuts)
- Waste of energy as a result of poor site selection (open, unprotected location)
- Negative effect on city- or townscape (by marring an existing harmony of architectural styles meriting preservation)
- Attracting automobile traffic (when not located near public transport stops or stations) with the consequent noise and exhaust generated by driving to and from the facility.

Building Design and Construction Process

- Paving of surfaces
- Waste of energy arising from poor configuration of component structures, or from improper orientation of facility to points of the compass; waste from inadequate façade and roof design, insufficient insulation of exterior walls, absence of heat recovery mechanisms in heating or ventilation systems
- Use of building materials that emit unhealthful gases, fibers, or dust into interior air (for example, wood varnishes, formaldehyde, asbestos), whose manufacture is energy-intensive or damaging to the environment, or whose proper disposal presents difficulties (for example, polyvinyl chloride plastics)
- Noise due to insufficient sound insulation of exterior walls and of heating and ventilation systems.

Operations

- Use of cleaning agents that contaminate water and degrade slowly
- Wasting energy (for example, in operating the heating system)
- Wasting water
- Use of non-recyclable materials and insufficient efforts to avoid generating trash
- Not sorting trash and collecting recyclable materials in separate containers
- Using materials and substances that are difficult to dispose of properly (generates hazardous waste)
- Air pollution as a result of choosing problematic sources of energy.

3. Finding Solutions

The following environmental criteria should inform and guide the planning, construction and operations of gymnasiums, indoor sports centers and indoor swimming pools (as well as of other buildings):

- Protecting areas that are both ecologically valuable and that play an important role in local climate
- Paving or covering surfaces only where absolutely necessary (areas covered by buildings, degree to which outdoor grounds are paved, for example, parking lots)
- Using environmentally-friendly means of energy and heating
- Choosing environmentally-conscious construction methods
- Using plantings on walls and roofs, and in surrounding outdoor areas
- Saving water
- Collecting rainwater or allowing rainwater to drain into the soil
- Using materials and substances not harmful to the environment or to human health
- Saving energy and using cleaning agents that do not harm the environment in the upkeep of facilities
- Avoiding activities producing noise.

The above goals can be achieved through measures focusing on the following spheres of activity (see also the chapter, "Protecting Natural Resources").

a) Design and construction of buildings

- Design a compact building in relation to neighboring buildings. Minimize the ratio between the exterior surface of the building and its interior space.
- Cut back on ancillary rooms and facilities for spectators.
- Passive use of solar energy:
  - Orient the building towards the south.
  - Use an appropriate design for the roof and façade.
  - Have the minimum amount of window space on the north side.
- Plant exterior walls (vines) and the roof.

b) Heating and energy technology

- Installing systems using renewable forms of energy
- Energy-efficient construction
  - High thermal storage capacity using thick exterior and interior walls
  - Favorable thermal conductivity constant in walls, windows, roof, and basement ceilings
- Energy-efficient heating
  - A controllable ventilation system and, where possible, a heating system using heat recovery from exhaust air
  - Low-temperature heating systems, ideally employing calorific technologies
  - Give priority to natural gas or centrally-generated heating. Electric heating should be avoided. As far as possible make use of new technologies (solar heating, combined thermal-electric power stations, heat pumps).
  - Minimize heat loss occurring during the production, distribution and control of heat (for example, in the installation of ducts in a building and by employing infinitely variable circulation pumps).
- Using recycled water for heating
- Electricity supply
  - Use energy-efficient electrical appliances.
  - Install energy-efficient lighting (low-energy bulbs).
  - When possible, use combined thermal-electric power stations.
  - When possible, use alternative sources of energy (solar, water, wind power).

c) Selecting building materials

- Choose materials whose extraction or production as well as processing entails the least amount of toxic emissions and use of energy.
- Use recycled materials as much as possible.
- Avoid building materials that would emit toxic substances into indoor air during and after construction.
- Choose materials which release a minimum amount of pollutants during construction and when disposed of after use. Avoid hazardous wastes (for problem-free disposal).
- Select materials that can be reused.
- Choose building materials that can be recycled after disposal.
- Select products and materials with a long life.

d) Environmentally-friendly construction procedures

- Plan how to minimize negative effects on the environment when tearing down or renovating buildings.
- Avoid releasing toxic substances into soil, groundwater, and air when building.
- Solving problems of waste disposal at construction sites:
  - Utilize all possibilities for recycling trash, construction materials, and construction debris.
  - Separate and properly dispose of hazardous wastes and other problematic substances.
- Avoid producing noise.

e) Building frame

- Construct exterior walls with energy-efficient materials. Sand-lime brick has favorable heat-conducting properties (thermal conductivity constant) and its manufacture consumes little energy.
- Materials and construction methods chosen for exterior insulation should be efficient and not pose health risks.
  - No styrofoam.
  - Use mineral fiber insulation materials without producing dust.
- In wood construction: use only preservatives that do not pose health risks.

f) Finishing touches

- Plantings on façades
- Greening of roofs
- For environmental reasons the following materials for interior sound and heat insulation are to be preferred:
  - insulating material made from cellulose or recycled material,
  - mineral wool free of fine particulates that invade into lungs,
  - wood fiberboard (also with gypsum content).
- Windows and doors
  - Material: wood whenever possible
  - No biocidal wood preservatives
  - Glazes, clear varnishes, weather-resistant paints, and lacquers should be water-based. Avoid organic solvents and other toxic substances.
  - Caulk around windows and make sure doors close tightly. When necessary, install windscreens.

- Non-supporting interior walls
  - Wood-chip board (not treated with formaldehyde) and plasterboard (particularly those made from recycled gypsum products) are recommended.
  - Pressboard made with synthetic resins should have at least an emissions class 1 rating (E 1).
  - Do not use building materials containing formaldehyde.
- Floors
  - Do not use synthetic flooring materials (they produce waste whose disposal is problematic) or those containing asbestos.
  - Use low-emission glues and filling materials.
  - Recommended flooring materials: wood (with surface coating, with lacquer coating made of natural resins), natural or artificial stone, ceramic tiles, linoleum, corkboard, and material made from natural fibers.
- Interior plaster and wall treatment
  - Products made from recycled materials are to be preferred.
  - Use water-soluble paints and lime-based paints.
- Use lacquers, glazes, and paints low in toxic substances.

g) Water supply and sewage system

- Install water-saving devices in toilets (levers that interrupt flushes), faucets and showerheads, particularly for saving warm water (faucets using a single handle for hot and cold water).
- Collect and use rainwater where potable water is not necessary (for flushing toilets, for example).
- Design toilet and shower facilities for easy cleaning (consequent decreased use of cleansers).

h) Exterior grounds (driveways, parking lots, courtyards, green strips and shrubbery)

- Minimize the paving or covering of surfaces or remove the pavement from them (except where the danger of pollutants seeping into the ground is present).
- Do not use pesticides and fertilizers.
- Preserve vegetation that has sprouted up naturally.
- Use native plants over others.
- Plant hedges and shrubs and keep those already extant (preferably untrimmed).

i) Facility operations and usage

- Schedule activities within facility to make the most efficient use of energy: activities. User groups that require the same or similar temperatures should be placed together.
- Reduce use of the gymnasium by classic outdoor sports.
- Adjust room temperature, ventilation, and lighting to the minimum levels required by a given sport.
- Keep water in swimming pools at the minimum temperature mandated by the KOK guidelines (24° C for indoor pools, 28° C for wading pools).
- The air temperature in halls with indoor pools should not exceed the water temperature by more than 2° C.
- Shut off circulation pumps for warm water at night.

j) Environmentally-friendly cleaning

- Recommended cleansers for use in sports halls and rooms, hallways, kitchens, toilet and shower facilities, and on windows and façades:
  - mild all-purpose cleansers and soaps,
  - vinegar (for removing lime deposits),
  - alcohol- and ethanol-based cleansers,
  - scouring powders.
- Use disinfectants in pools that do not harm the environment (instead of chlorine).
- When cleaning façades: catch all water flowing down from the façade and dispose of it properly.
- Use cleaning appliances and devices that require little energy, water or cleanser.
- Do not use air fresheners or deodorizing sprays in lavatories, or place solid disinfectants in toilet bowls.
- Supply toilet paper and paper towels made from recycled material.

TENNIS (F)

1. Significance to the Environment

Tennis is played in summer on outdoor courts and in indoor halls in the winter. A tennis center usually consists of at least 2 to 3 courts with an average of 6. It includes a clubhouse with locker rooms and showers, and in especially large centers, grandstands are also found. Some clubs have indoor courts (usually 2 to 6 in number).

The courts themselves usually comprise 50% of a center's total area. When the courts are indoors the percentage is higher. According to the German Tennis Federation (DTB) the other 50% is taken up in equal measure by the clubhouse, parking lots, pathways, gradients and green spaces.

Loudspeakers are used chiefly during tournaments. In many cases flood lighting is part of the facility's equipment. The following infrastructure also forms a part of tennis centers: driveways and walkways, infrastructure for power and water and for waste disposal (sewage and solid waste), as well as parking lots.

2. Potential Conflicts and Ways to Solve Them

In tennis the principal problem is noise in the vicinity of residential areas. The greatly increased number of tennis players and the lengthened hours that tennis centers stay open (people now play during normally quiet times such as mornings, weekday evenings, weekends and holidays), combined with the increased sensitivity of residents to noise and sounds, have given rise to conflicts. Some of these have even led to court battles. These conflicts in turn often rest upon such mistakes in planning as failing to leave enough open space between the tennis center and adjoining neighborhoods or to employ other measures for reducing annoying sounds.

Noeke and Rolf (1986) have explained the particularly annoying features of tennis-related noise as follows: "Tennis-related noise entails a particular unpredictability. The sounds of a tennis ball being hit back and forth has a unique rhythm that even a person who merely hears but does not see the activity can quickly recognize: the famous plop-plop sound. Yet his rhythm is often abruptly broken, when, for example, a ball lands in the net. This sudden cessation of sound creates a tension in the hearer, for it was unexpected. An expectation of hearing a sound is thus continually reinforced and interrupted by this kind of irregular regularity of sounds."

The illustration below shows a protractor-like model developed by Probst for estimating the intensity of sound received in a residential area adjacent to a tennis court. Intensity of sound varies according to distance from the court.

We list below in brief outline some ways of solving the problems caused by noise issuing from tennis courts (see also chapter B).

Decreasing sports-generated sounds and noise through technological and organizational means:

- The sound produced from tennis balls striking walls can be softened by shielding the walls with nets or foam matting.
- Loudspeakers distributed throughout the facility and aimed at small groups generate less noise than a centralized loudspeaker system.

- Clay courts are quieter than paved ones.
- Noise absorption measures: planning, plantings, berms and walls (more on this below).

When all other measures are insufficient to solve the problem, playing times must be limited (imposing quiet times).

Such technological means of reducing noise as flexible sound-insulating tarpaulins (a three-dimensional covering filled with sand) and planted sound berms (wickerwork laid over an earthen core) have proven effective. Coverings with a weight of 20 kg/m2 have decreased sound intensity up to 38 dB(A), while planted berms allow a decrease in sound intensity of up to 30 dB(A) at a distance of 5 m away.

The great amount of **paved over surfaces** constitutes another impact on the environment. It thus becomes all the more important to leave the few uncovered green areas in as close to a natural state as possible. Green spaces should be mowed infrequently (twice a year is sufficient) and, where possible, left to grow and develop on their own without any outside interference.

Opportunities for recycling space (for example, using former industrial zones) should be utilized.

In order to minimize **consumption of resources,** environmentally-friendly sports clothing and equipment should be selected. Measures to conserve resources, such as saving energy and water, should also be implemented in clubhouses (see chapter A).

Disturbance caused by **exterior lighting** can be avoided when flood lighting is positioned not to shine into adjacent neighborhoods. The same regulations that impose quiet times in order to eliminate annoyance from noise (for example from 8:00 to 10:00 p.m.) also serve to prevent lighting from disturbing nearby residents. When siting illuminated tennis courts on the edges of urban areas, keep in mind that such facilities act as an ecological "trap" on insects (see chapter A).

SHOOTING AND ARCHERY (G)

1. Significance for the Environment

The term shooting and archery comprises the following sports: rifle shooting, pistol shooting, skeet shooting, bobbing target shooting, archery, crossbow archery, shooting muzzle-loaders.

Three types of shooting sports are relevant for the environment:

a) Rifle and pistol shooting (including with air guns and CO2 guns)
b) Skeet shooting (shooting "clay pigeons")
c) Archery using crossbows and regular bows.

Because archery produces little noise, a flat meadow equipped with the requisite security features is an appropriate site for it. Shooting with firearms, however, because of the noise it produces, requires buildings that are open to the outside, partially roofed, or completely enclosed.

A range house normally shelters the individual shooting stations or boxes, from which patrons shoot outward into the air. For safety reasons the shooting ranges must have side barriers (berms and walls), high protective blinders, and range boundary markers. The shot projectiles are collected in a trap. Roofed and enclosed shooting boxes are the norm in shooting with air guns.

In skeet shooting a moving target is fired at with small-shot charges. A skeet range of three hectares in size completely lacks barriers and consists simply of a flattened out field. The clay pigeons are hurled from two towers into the range of fire. In trapshooting the targets are flung upward from a trench. The spent lead shot falls to the ground over a large area.

2. Potential Conflicts

Shooting ranges damage the environment principally through noise and causing the entry of harmful substances into the soil (lead contamination).

a) Noise

Because of the nature of the sound, noise from shooting can cause significant stress on people living near ranges.

Depending on sound reduction measures in effect, open shooting boxes cause sound intensities of at least 50 dB(A) in an area 80 to 300 m distant. Noise from skeet shooting is particularly loud because the shooting range cannot to any significant degree be enclosed, either with walls or a roof.

Sound intensities measured in A and the energy-equivalent average value are fundamentally suited for measuring and evaluating noise from shooting firearms. Noises from shooting, however, are considerably more irritating than others having the same average sound intensity LAm, such as street noises (UBA 1989). In shooting the venue is a decisive factor. It makes a great difference if the shooting of firearms takes place indoors, such as in roofed ranges or underground cellars, or in outdoor fields where the lack of barriers allows sound to penetrate much further into the surrounding environment.

A judgement as to whether a given level of shooting noise is acceptable must be made according to the legal limits set for sound intensities: does the shooting noise in a particular case fall above or below these values? The figure used in reaching this decision is a composite value based upon the sound intensity of individual shots, the number of shots fired within a space of time, the number of hours per day and the time of day during which shots are fired. Explosions from the discharge of large caliber ammunition from the muzzle, and from shots exceeding the sound barrier, cause the greatest sound intensities.

b) Lead Contamination

Skeet shooting using lead pellets in particular contributes to the problem of contamination of soil and groundwater by heavy metals. Up to one million shots a year are fired in a skeet range. This number can be much lower in smaller ranges (for example, only 20,000 shots). This means that up to 30 tons of heavy metals a year can fall on the ground in a skeet range (a cartridge contains at the most 32 gr. of shot).[4] Depending on the number of years it has been used, a shooting range three hectares in size could thus harbor several hundred tons of lead. The suspicion that lead deposits have accumulated in some areas thus may not be casually dismissed.

A study that examined concentrations of lead, arsenic, and antimony in the soil of shooting ranges, and the long-term environmental effects of the absorption of these metals by plants, is of particular interest here. It showed that lead content sometimes exceeds legally prescribed limits, and that this poses a consequent risk for agricultural products and groundwater (UBA 1989).

4 The shot cartridge varies from between 24 and 32 g in size. For shooting ranges 24 g is prescribed.

3. Finding Solutions

Avoiding Contamination by Lead

The following guidelines should be observed in order to prevent serious contamination of soil and groundwater (see also UBA 1989):

- To protect nature and water supplies, shooting ranges should not be permitted in protected watersheds that provide drinking water, nor in groundwater catchment areas used for the public water supply. Lands with a high groundwater level or high water permeability, in nature reserves, and near open bodies of water are also unsuited for ranges.
- Land near shooting ranges may not be used for agriculture or grazing of animals.
- Nickel-coated shot should no longer be used.
- Soils in shooting ranges should have a high absorption capacity. The application of humus can help achieve this. Soils whose content of arsenic and antimony exceeds legal standards should be treated to obtain a pH value between 4.5 and 5.5. This will help to combat the movement of heavy metals downward through the soil.

Because heavy metals are so harmful we must ensure that

- only ammunition that is low in lead or lead-free is used,
- as much lead shot as possible is collected by shot gathering equipment,
- ranges are kept a sufficient distance away from surface bodies of water, from protected watersheds, and from groundwater lying near the surface, so that contamination of surface and groundwater does not occur,
- shot does not fall on areas accessible to wild animals (birds, mammals) with the result that animals are poisoned from eating food contaminated with lead,
- the soil in and around older ranges is tested for lead content,
- when a skeet range is permanently closed all soil that could contain lead bullets is dug up to a depth of 20 cm and removed (toxic waste).

Further recommendations for shooting ranges concerning shot:

- When the range sits on land having a slope greater than two degrees and consisting mostly of water-impervious soils, care should be taken that soil containing heavy metals does not wash off onto neighboring agricultural land or into surface bodies of water. Thick vegetation cover can prevent runoff in such cases.

- Cartridge cases should be picked up after each shooting session and properly disposed of. Whole clay pigeons and broken chunks of them should be gathered and disposed of as garbage. (The grass where pigeons usually fall should be kept short in order to make the collection of pigeons easier.)
- Shot with nickel coating may not be used.
- Paints containing heavy metals may not be used on pigeons.
- The number of pigeons thrown each year should be recorded, for example, on the basis of purchase receipts. Receipts should be saved.

Avoiding Noise

The following measures can protect neighboring areas that are sensitive to noise (residential areas, recreational areas, medical care facilities):

- Roof over the facility.
- Maintain sufficient open space between the shooting range and sensitive neighboring areas.
- Construct sound berms or walls.
  - Limit the hours when shooting takes place (pauses during quiet times).
  - Use subsonic ammunition (when possible).

WALKING, HIKING, AND RUNNING (H)

1. Significance to the Environment

This chapter deals with hiking, walking, and jogging and the infrastructure required by these activities (in the main paths and mountain cabins). Hiking and walking is a sports discipline without the element of competition in which people either individually or in groups cover a distance ranging from 5 to 50 km a day. Hiking and walking comprise a variety of paces and exertion levels, from leisurely strolling to hiking over difficult terrain, from competitive walking (a sport practiced over short distances) to trekking and long-distance hiking. Those who engage in walking and hiking as a form of recreation, rather than as a strenuous sport, look to practice their activity in attractive scenery. Especially preferred are well-forested areas (with open, grassy valleys), areas featuring varied topography (medium and high mountain ranges) and bodies of water (streams, lakes, river valleys, seacoasts). Closeness to nature is desired here. Forests with a mixture of trees, for example, are preferred over monotonous plantings of spruce, streams having natural riparian zones are sought out over streams with graded, paved

banks. When available, people prefer to walk on unpaved trails and paths as well as on unpaved farm and forest roads. Asphalt, though regarded as annoying, is accepted out of necessity.

Hiking and walking impinge on the environment chiefly through the laying out of trails (their courses and density within a given area) – unless farm and forest roads are used – and through the construction of other infrastructure. Hiking-related infrastructure such as parking lots at trail heads, picnic and recreational grounds with barbecuing pits, shelters, interpretive and exercise trails, bridges, trail signs and markers, informational and educational boards, and mountain cabins also affect the environment.

Running may be practiced on one's own as a form of recreation or in association with a club as either a popular or competitive sport. As do hikers and walkers, runners prefer attractive scenery for their activity, though for runners such landscapes (particularly those in a near natural state) are less important. Runners use both unpaved and paved farm and forest roads as well as regular streets and roads.

A special variation on running is the **orienteering run** carried out according to established rules. In this contest participants must pass by a series of checkpoints using only a compass and map for help. The course, up to 16 km in length, must be traversed in the shortest time possible. The courses lie often in forested areas, 90% of the time away from established paths and trails.

2. Potential Conflicts

Hiking and the infrastructure it uses can have the following negative impacts on the environment:

a) Stress resulting from trail construction:

- when trails are surveyed and laid out without paying attention to the local environment: loss of valuable habitat from paving trails or from using fill to construct trails through wetlands (with consequent negative effects on drainage patterns),
- when soil or material used as substratum for trails is imported from elsewhere (for example, using limestone in an area with acid soil): vegetation growing along the trail is then placed under stress.

On the other hand trails also relieve stress on the environment by channeling people into ecologically unproblematic areas.

b) Stress arising from people leaving marked trails (not observing rules prohibiting this):

- Disturbance of protected, shy animals
- Water erosion on steep slopes (caused by people cutting across corners in winding mountain trails).

The above listing should make clear that leaving trails only constitutes an environmental stress when it takes place in a protected or valuable area (where visitors are specifically directed to stay on trails), such as moorlands. Not every area is so ecologically sensitive.

Ecological stress such as the destruction of vegetation from trampling is not limited, however, to protected areas. It can occur in any small area where great numbers of visitors congregate or pass by (for example, near certain mountain peaks or in other similarly attractive points in the landscape).

c) Stress arising from outdoor events:

- Disturbance of animal communities when orienteering runs are conducted in sensitive areas and during critical times (for example, breeding season)
- Problems caused by people driving to and from an event (noise and air pollution) and from parking during events (destruction of vegetation in near natural areas).

d) Stress caused by inadequately handling waste from mountain cabins, mountain lodges, hotels, and restaurants, as well as restaurants located near glaciers (leading to contamination of groundwater and surface water):

- Insufficiently treated sewage
- Collecting garbage into outdoor piles or dumps (leakage of contaminants, marring the visual beauty of the landscape).

3. Finding Solutions

a) Laying Out Trails, Hikers' Behavior

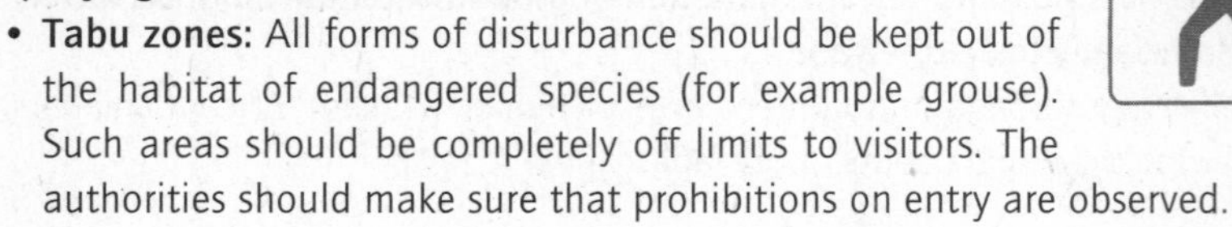

- **Tabu zones:** All forms of disturbance should be kept out of the habitat of endangered species (for example grouse). Such areas should be completely off limits to visitors. The authorities should make sure that prohibitions on entry are observed.

- **Not providing trails:** Plant communities in particular need of protection (such as moors, swamps, dunes) should be completely free of trails. In exceptional cases raised wooden walkways are permissible from an ecological point of view (they do not divide or cut off areas, and cause a minimum of disturbance).

- **Directions to stay on trails:** A prohibition against leaving trails should be enacted and enforced when necessary in nature reserves and other valuable and sensitive areas. The network of trails in such areas should be kept to a minimum. Prohibitions against leaving trails are only justified from an ecological point of view when valuable plant communities or animal populations would be endangered, the former by trampling, the latter by other disturbances.

- **Removal of trails:** When existing trails pass through particularly sensitive protected areas and prohibitions against leaving trails are continually ignored, the affected trails should either be closed or be removed.

- **Organization of large outdoor events:** Orienteering runs and similar large-scale events in forests and other near-natural areas should only take place after consultation with the nature conservation authorities in charge of the area. In this way events can avoid sensitive areas and times.

- **Limiting automobile access:** The natural areas closed to vehicular traffic should be as large as possible. Roadways leading into highly frequented recreational areas should, whenever possible, be closed to private automobile traffic, and the transport of hikers accomplished instead by horse-drawn wagons or public transportation. A small number of parking spaces should be allowed on the edge of such a recreational area, which only pedestrians may enter. These parking lots should be well integrated into the landscape (no asphalt).

- **Unpaved trails:** As a rule of thumb paths and trails should be paved as little as possible, and in any case never with asphalt. This will keep the division of ecosystems through the "cut-off effect" to a minimum. Hikers can be expected to wear water-resistant footwear. Walking on unpaved trails also has beneficial health effects.

- **Sponsoring:** Utilize opportunities to involve hiking clubs in protecting biotopes or restoring them to their natural state. Hiking clubs can act as sponsors for protected components of the landscape, natural monuments, nature reserves and fallow agricultural land.

- **Managing visitors:** By means of "soft" (because unnoticed) channeling, visitors are guided to trails leading to areas and points in the landscape that are simultaneously attractive and able to withstand stress. "Psychological barriers," such as plantings, water channels, piles of branches, or wooden planks lining the trail, can also be used to direct the movement of visitors.

- **Taking hunting interests into account:** Disturbance of legal game animals is not an ecological stress, but a conflict with hunters' interests. Measures to limit conflict (for example, respecting wildlife reserves, not hiking at nighttime, avoiding meadows populated by game animals, rutting grounds, and places where feed is left for game) can be negotiated between hiking clubs and hunters. Negotiations to shorten the hunting season and over introducing hunting methods that make game animals less shy (such as certain forms of battue or shooting game in winter corrals) should form a part of this process.

- **Information and consciousness-raising:** Hikers should receive more information on ecologically respectful forms of behavior. Such information could, for example, describe how to dispose of trash, how to choose a resting spot, and how to take dogs along on hikes.

The rather restrictive measures outlined here are not, however, the only means for finding solutions. Hikers' love for the undisturbed landscape should also be encouraged and fostered. This love shows itself not in placing ill-considered claims and demands on nature, but in respecting its ecological limits and sensitive points.

b) Mountain Cabins

By following the guidelines outlined below, the operators of mountain cabins can help keep stress on the environment to an absolute minimum:

- The capacity of existing cabins should not be expanded (for example, to accommodate larger numbers of visitors during peak seasons).

- No increases in comfort levels such as having cold and warm running water in the rooms or offering more sophisticated cuisine (rather, the comfort level should be reduced).

- Only local products and dishes typical of the region should be offered.

- Set up self-contained supply and disposal systems (a closed circular system).

- Use adapted technology for all supply and disposal systems (use of sun, wind and water for energy; use surplus energy, for example, to treat sewage more efficiently).

- Replace diesel and gasoline-powered generators and gas lighting as far as possible (this will prevent noise and odors, emissions, and contamination of drinking water through seepage of diesel oil into groundwater). If these generators cannot be replaced, at least link heat and power systems to increase efficiency.

- Install sewage-treatment facilities that deliver a high level of purification (using state of the art technology whenever replacing equipment or installing new equipment).

- Proper disposal of trash and waste (priorities are avoiding, reducing, and recycling).

- Use environmentally-friendly sources of energy and use them sparingly; avoid using drinking water when possible.

ROCK CLIMBING AND MOUNTAINEERING (I)

1. Significance for the Environment

It is relevant for the environment to distinguish between climbers active in high (alpine) mountain regions and those in mountain ranges of lower elevation. Rock climbing in lower-elevation highlands, such as the Central German Uplands, can lead to serious conflicts with the goals of nature conservation because the rocks suitable for climbing are scarce (and the number of climbers relatively high). In high mountain ranges (such as the Alps) climbing causes scarcely any conflicts worth mentioning.

Areas of rock occur relatively often in the Alps below the timberline and abundantly above it. Ecosystems typical of the Alps have developed in these rock areas. In mountains of lower elevation, however, the dominant ecosystem is forest. Rock outcroppings appear here only in a few scattered locations. They are, moreover, special areas that provide a habitat for plant and animals species of great ecological significance. Often these are isolated animal and plant communities without any genetic exchange with other communities ("disjunctive areas" and relict habitats). The discussion of conflicts and possible solutions in this manual will thus focus on lower-elevation mountain ranges, the Central German Uplands in particular.

Rock climbing in such highlands is almost always free climbing,[5] that is climbing without any technical aids along short, mostly well-secured routes, whose starting points are easily and quickly accessible. Several rocks that would be suitable and interesting for climbing have had to remain closed to this sport for reasons of nature conservation.

Canyoning is a sport now much in vogue that lies somewhere between rock climbing and water sports. This new sport originated in America, but also in Spain and France, and has since spread to other countries.[6] Canyoning has special significance for the environment since it takes place in natural areas that up till now have been largely untouched by man and consequently have a pristine, primeval character to them. It remains to be seen if this new sport will have a negative impact on the undisturbed plant and animal communities in these areas.

---

5 "Free climbing is understood as movement using only the rock's natural hand- and toeholds, that is without weighing down the safety rope between the start and exit points of the climb or between the individual resting points along the climb." (Hoffmann/Pohl 1996)

6 "The popularity of canyoning is booming, this is everywhere to be seen. ... The French alone registered 50,000 climbs in gorges last year." Canyoning is on the way to becoming a national sport in France (Bergwelt ALPIN 8/96).

2. Rock Outcroppings in Mountain Ranges of Lower Elevation: Worthy of Protection

For centuries the biotopes of the rock outcroppings in low mountain ranges were relatively free of human interference. Thus primeval, natural habitats with a rich variety of animal and plant communities, which scarcely occur in an intensely farmed and cultivated landscape, have survived here. The rock cliffs exhibit extreme conditions in comparison to those in their surrounding areas (wide swings in temperature, effects from wind, lack of nutrients, high level of solar exposure). Animals and plants adapted to such conditions are "specialists" that very often appear in the Red List of endangered species. Areas of exposed rock along with a few other biotopes (bogs, larger reed beds, natural alluvial ecosystems along rivers) are the last undisturbed primeval habitats. Rock outcroppings and their immediate surroundings constitute the only places where many species survive since the ice age.

The following areas of rocks found in low mountain ranges are susceptible to environmental stress from climbing, or require special protection from stress brought on by rock climbers:

Rocky knobs: The varied and highly specialized vegetation and fauna on rocky knobs is particularly sensitive to trampling or disturbance. Peregrine falcons, for example, use these rock formations for perching, consuming prey, and sleeping. Rocky knobs are the most sensitive of the areas visited by rock climbers. They are "exit points" for climbs up rock walls or the peaks of freestanding pinnacles.

Walls: Rock walls with cracks, crevasses and holes along with rock ledges also shelter a rich flora and fauna. Vegetation growing on these formations is particularly susceptible to trampling. The animal communities here are sensitive to outside disturbance as well. Peregrine falcons and ravens nest in niches in the rock.

Walls that are covered with vegetation are not as interesting for rock climbing as are steep, often overhanging rocks without vegetation. For it belongs to the nature of this sport that climbers seek out higher levels of difficulty.

Entry and Exit Points for Climbs, Bases of Rock Walls: The plant communities in these zones (scree and boulder piles), which provide a home to small animals such as insects and reptiles, differ from those found on the shady forest floor. They require warmth and sun, and are easily damaged by trampling and erosion.

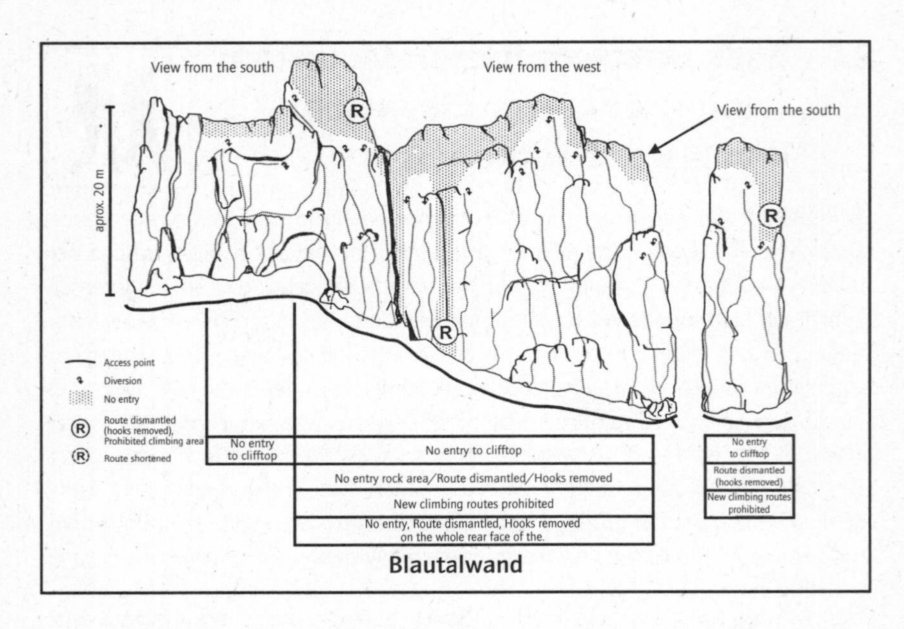

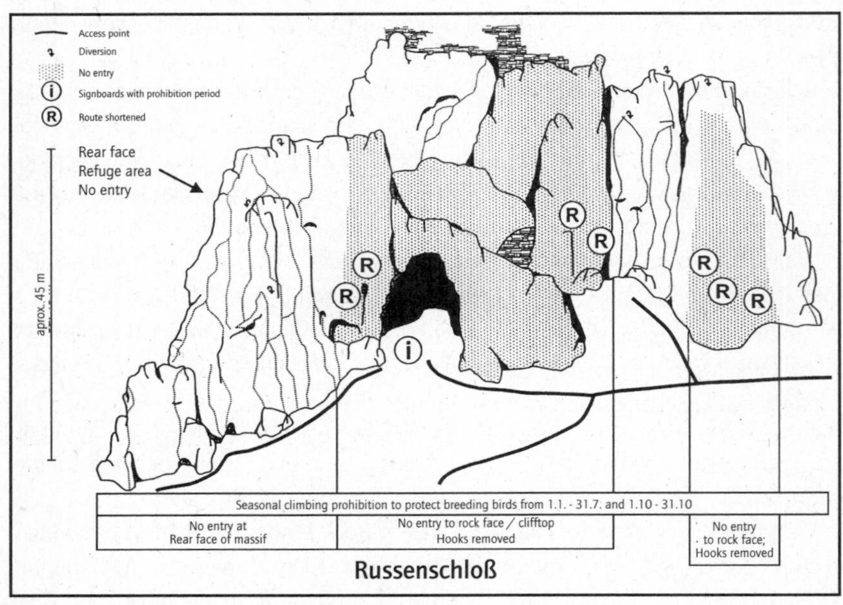

Source: KREH et al. (1999), Graphic D. Scheuer

3. Potential Conflicts

The enormous increase in rock climbers in the Central German Uplands poses an acute or a potential danger to the valuable qualities found in their rock biotopes. Climbers' present widespread use of rocks that they previously considered "uninteresting" brings greater risks for the sensitive animal and plant communities of rock biotopes. The increase in the number of climbers and the expansion of the times used for climbing have also contributed to this problem.

Rock climbing and activities related to it[7] have had the following effects on animal and plant communities in rock biotopes (see Bichlmeier, 1991; Senn, 1995; Ficht et al., 1995):

Negative effects on **valuable species of plants and small animals** resulting from

- **Trampling:** Such direct mechanical impact pushes out species sensitive to crushing. Trampling can negatively affect entire habitats over the long-term: less on the actual rock wall itself, rather in those areas immediately around the points where climbers enter upon a climb and exit from it.
- **Depositing of substances:** Inorganic and organic wastes (including excrement and ashes from campfires) cause eutrophication of the soil. As a consequence, rare plant species are displaced by more adaptable common species.
- **Removal of material:** The intentional removal of organic and inorganic substances when setting up for a climb (such as clearing vegetation out of the way).

Negative effects on **valuable species of larger animals** resulting from:

- **Outside disturbance** (normally concentrated and lasting for short periods of time.) is a problem mainly during the breeding period for peregrine falcons, eagle owls, jackdaws and ravens. Reptiles living in the rock [such as smooth snakes (Coronella austriaca) and wall lizards (Podarcis muralis)] and bats are also highly susceptible to human-caused disturbance.
- **Introducing changes into biotope structure:** through removal of vegetation.

---

7 Driving to and from the site, hiking up to the base of the rock, camping, climbing, changing routes in the course of the climb, exit from the climb

4. Finding Solutions

In Germany's Central Uplands an increasing number of rock climbers meets with a strictly limited number of rocks suitable for climbing. Since the number of natural rocks cannot increase, the following measures become necessary.

a) Fully utilize (with respect to space and time) those rocks that can be climbed without causing significant environmental damage.
b) Prohibit any climbing on rocks other than those described above, or direct would-be climbers to artificial rock walls.

Attempts are presently under way to determine which rocks are suitable for climbing. Here the task is to find out which rocks may be scaled without causing serious harm to the animal and plant communities living on and around them. **Any master plan for climbing** must rank areas of rocky outcroppings and individual pinnacles according to their ability to withstand the effects of climbing:

- Which rocks must remain entirely off-limits to rock climbers (total prohibition all year round)?
- Which rocks may be climbed at least in certain parts or at certain times without placing the environment under serious stress (prohibiting climbing during certain seasons, setting some rocks off-limits or closing certain climbing routes, visitor management measures)?
- Which rocks may be climbed only during certain times or only in certain areas? Here it is sufficient to ensure that activities accord with the standards of "soft, environmentally responsible" rock climbing (for example, no clearing of vegetation).

Use plans should be created for all rock climbing areas. Plans should differentiate areas according to the level of use that they can withstand and strive to make the interests of climbers compatible with conservation goals.

The following essential points require attention when formulating a use plan for rock areas:

- **Ensuring that climbing will not harm the natural balance of an area:** Rocky outcroppings are of great ecological importance.[8] In addition, they often constitute important parts of biological corridors. For this reason a use plan must make sure that climbing activities do not damage the valuable ecological qualities of rock areas. Sports enthusiasts and conservationists often disagree about which changes introduced in nature are to be regarded as "significant". Thus environmental studies must be conducted with care, and any protection measures or policies arising from them have a firm, scientific basis. In the event that the two sides draw different conclusions from the data, the authority responsible for nature conservation in the affected area shall have the last word.

- **Input from rock climbers** in the formulation of a master plan for nature conservation and climbing: Giving climbers a voice in the planning process makes an objective and fair solution possible because climbers' interests can be worked into the final regulations before they are issued. This is important for the regulations to win acceptance and makes their implementation easier.

- **Making maximum use of areas eligible for climbing:** The possibilities to carry out rock climbing in harmony with nature cannot be described according to some general scheme. Rather, each rock area must be carefully evaluated on site. The various natural conditions found in a given rock area (such as the quality of habitat for certain species, the degree to which the rock lies exposed) must be considered here, as must the ecological value of the area, its accessibility to climbers, and the type of climbing proposed in it. The boundary separating zones where climbing is permitted, and where it is strictly off-limits, must be drawn and marked clearly and unmistakably: rock outcroppings and portions of them that are open to climbers must be so designated, all other rock formations in the area are closed to climbing.

- **Channeling climbers by means of trails:** In order to steer people away from ecologically problematic rock areas, roads leading to them in the surrounding area should be closed at a good distance away. This will at least keep those climbers wanting to avoid an arduous hike away from such sites.

---

8 Exposed rock outcroppings in Germany's Central Uplands belong to those biotopes protected by §20c of Germany's Federal Nature Conservation Law.

- Information and education: Educational efforts must aim at bringing climbers to think of a rock formation not just as a sports site, but also as the habitat of animals and plants worthy of protection. Climbers' understanding of the site as habitat thus constitutes the prerequisite for appropriate behavior in nature. Guidebooks for rock climbing must not undermine the goals of nature conservation, but should encourage respect for them by describing only rocks where climbing is permitted.

- Cutting back on climbing routes: The removal of existing climbing hooks on rocks that for environmental reasons must be closed to climbing should be accompanied by information explaining to climbers the reasons for the closure.

- Providing attractive alternatives that are ecologically unproblematic: Rock areas able to withstand environmental stress and thus requiring scarcely any limitations on climbing, as well as artificial rock walls, both belong in this category. At present, man-made walls for climbing – erected either outdoors or inside gymnasiums – are becoming increasingly common.

- Restoring rocks used for climbing: By "restoring" we mean caring for rock areas and developing them in a way that removes or lessens negative ecological and aesthetic effects on them. This term also includes measures to prevent such negative effects from occurring in the first place.[9]

Gorges that come into consideration as sites for canyoning must first be carefully studied to see whether they are suitable for such an activity. In the event that they are, plans must be developed for managing visitors.

---

9 The usual understanding of the term "restoration" focuses on the removal of safety risks (such as the replacement of hooks pounded into the rock with ones that are drilled into it, the kind normally used today). As part of its "Project on Mountaineering and the Environment," the German Alpinist Association (DAV) has issued a "Plan for the Restoration of Non-Alpine Rock Climbing Areas" for its local chapters and working groups that supports the measures discussed here.

MOUNTAIN-BIKING (J)

1. Significance for the Environment

Since the 1970s cycling has experienced a boom in Europe, certainly a great progressive development from an environmental point of view. As an alternative to motorized vehicles the bicycle is rightly thought of as the environmentally conscious choice of transportation.

Other than the conventional bicycle, the mountain bike is designed for cross-country riding.[10] The sturdy materials used in the mountain bike enable it to be ridden off-road, making it a potential agent of stress on the environment. Because of this off-road capability and the relatively large place occupied by the mountain bike in present discussions on sport and the environment, this handbook will not deal with the sport of cycling in general, but will instead focus on the use of mountain bikes.

People ride mountain bikes as part of everyday commuting, for pleasure, as a popular sports activity, and in sports competitions. The various types of bicycles (city bikes, all-round bikes, fun bikes, and racing bikes)[11] differ in the stability of their frames, in their equipment, and in the thickness and traction of their tires. They differ correspondingly in their suitability for off-road use. Only mountain bikes from the "fun" and "racing" categories can be ridden satisfactorily off-road. Mountain bikes are used in the following sports activities:

- Cross-country (CC) racing: The racecourse takes contestants through a variety of terrain on dirt and gravel paths, through forests and open fields, and on paved roads. The percentage of paved roadways in a course may not exceed 15%.
- Downhill racing (DH).
- Hill climbs (HC) or uphill racing.

In addition there are parallel slaloms, long distance races (marathons), sprint contests, and contests testing dexterity (trials). In a few cases cross-country races and dexterity trials are also held in indoor halls.

---

10 This distinction does not always hold, however, since many people use mountain bikes as everyday bicycles.

11 The "trekking bike" is not classified as a mountain bike. It has larger tires (28 in as opposed to 26 in) and is often described as "a mountain bike for flat terrain."

2. Potential Conflicts

Mountain biking can give rise to social conflicts in addition to ecological ones. That the social conflicts brought on by the mountain bike – the ones in the forefront of public debate – are sometimes called "environmental problems" results from setting imprecise, fluid boundaries on the term "environment." The present handbook will not discuss conflicts with walkers, joggers, horseback riders or hunters, for these do not touch on the "environment" as the term is here understood.[12] These are rather conflicts arising between groups seeking to use the same space (usually trails) at the same time and getting in each other's way in the process.

Mountain bikers can cause **ecological** stress

a) when they leave paths and trails[13] and (as a consequence of brake marks and other mechanical effects) thereby damage vegetative ground cover (such as turf) which in turn may cause erosion to develop,

b) when, riding on trails or off-road, they travel through ecologically valuable habitats of sensitive animal species,

c) when in the course of large outdoor events (competitions) soil and vegetation are harmed (from trampling) and trash is not removed,

d) when they use cars to travel to and from the place where they will ride, thereby producing noise and air pollution.

Various estimates are available of the percentage of **mountain bikers who leave paths** when cycling. In a survey conducted by Weigand (1993) in the Taunus region of Germany, 14% of the 1,000 mountain bikers surveyed[14] said that they frequently ride cross-country. Analysis of several other surveys (cited in Wöhrstein, 1993) has shown that the percentage of mountain bikers who ride cross-country lies between 1.5% and 16%, whereby the number of off-road bikers in recreational areas near metropolitan centers is much higher than this average.

---

12 "Environment" does not refer here to "social environment," but to the physical environment as the object of environmental protection and nature conservation (soil, water, air, fauna and flora, scenic qualities).

13 Damage accruing to trails themselves may not be regarded as a stress on the environment.

14 In comparison, 27% of walkers said they frequently leave trails.

Damage to vegetation and soil caused by off-road mountain bike riding is only serious when it occurs in natural areas with valuable animal and plant species, or when it gives rise to significant erosion.

Mountain bikers cause the most serious stress on the environment, however, when they move into remote, previously "untouched" areas and so drive animals belonging to rare and sensitive species out of their habitat. Such activity that places stress, for example, on species of grouse (populations of capercaillie, black grouse, hazel grouse, and snow grouse) in their already too small remaining biotopes is simply indefensible.

Twiehaus (1994) studied twenty selected **bicycle races** held in Switzerland, Austria, and Germany. On average between 500 and 1,000 people participated in these events. Some events, however, had over 20,000 spectators in attendance. The author reaches the following conclusion: "Damage to the environment caused by mountain bike races is relatively minor. Only at events with more than 2,000 spectators does damage to soil, flora and fauna take on a greater scale, although even then it stays within tolerable limits." Because the bicycles have such a light weight, the root systems of plants are normally not damaged. Thus grassland vegetation occurring along the racing route is able to recover within a single growing season.

3. Finding Solutions

By far the majority of conflicts caused by mountain bikers are of a social nature (conflict with people on paths and trails). The recommendations below concern only the relatively minor ecological problems arising from mountain biking.

- **Protecting sensitive habitats:** Trails used by hikers and bikers that lead through habitats of valuable and sensitive animal species should either be closed or rerouted. In some cases – in consultation with the local conservation authorities – such problematic trails need only be closed during certain times of the year.

- **Providing attractive biking routes:** The creation of a network of interesting bike paths (along with the provision of adequate information on it) can direct mountain bikers away from sensitive areas. People will also avoid narrow erosion-prone trails when better alternative routes are available. Here is where planners need to get involved. Mountain bikers should be consulted when setting up such a system of marked paths.

- **Environmentally-responsible racing events:** The site and route for a race should be chosen according to criteria that will preclude contestants and spectators from putting valuable animal and plant habitats under stress.

- **Keeping bikers on allowed trails:** In order to hold down the "temptation" of off-road riding as far as possible, people should not be allowed to take bikes with them on lifts or cable cars going up into the mountains. In this way bikers can be kept from riding over mountain pastures as they descend into valleys. Also such a measure reduces the number of bikers who could cause disturbance in remote and valuable animal habitats.

- **Education and appeals for cooperation:** Rules for conduct should inform mountain bikers about the environmental problems associated with riding and the reasons for prohibiting off-road riding. Bikers should also be requested to use public transportation to take their bikes to outdoor sites.

Advertising has an important influence on mountain bikers' behavior. Advertising that does not propagate the "adventure" of riding off-road through natural areas would go a long way toward reducing conflicts arising from biking off trails and roads Biking associations should work with manufacturers to change such advertising messages.

EQUESTRIAN SPORT (K)

1. Environmental Significance

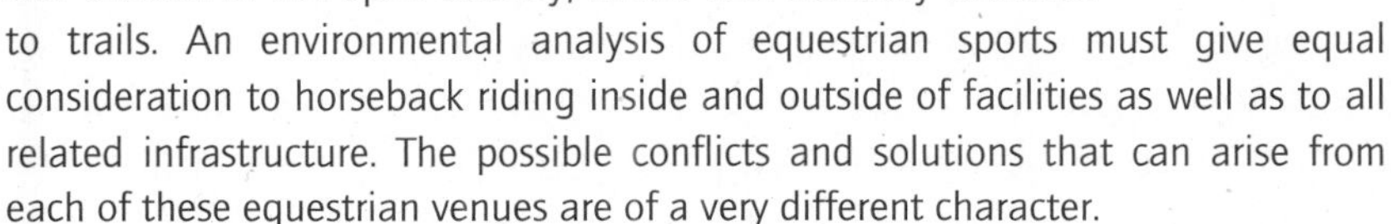

Horsemanship takes place both in facilities (halls and stadiums) and outside in the open country, where it is normally confined to trails. An environmental analysis of equestrian sports must give equal consideration to horseback riding inside and outside of facilities as well as to all related infrastructure. The possible conflicts and solutions that can arise from each of these equestrian venues are of a very different character.

This chapter will consider only horsemanship as a popular sport and the competitions connected with it, since most of the effects of equestrian sport on the environment stem from these activities.

Established equestrian disciplines (as parts of a popular sport):

Dressage:	training the horse to perform movements of harmonious precision
Show jumping:	jumping over barricades
Versatility:	a combination of dressage, show jumping and endurance competition
Vaulting:	performing acrobatics on horseback
Driving:	hitching horses to carriages or wagons
Distance riding:	covering long distances on horseback
	Horseback riding at a slow gait through the landscape as a form of recreation, either as a short excursion or a trip lasting several days.

2. Potential Conflicts

The following list of conflicts that can arise from equestrian sport considers only those involving the environment. Conflicts with groups using the same areas as horseback riders (such as farmers, foresters, motorists, hunters, and other people seeking recreation) have been intentionally excluded. So has the most common conflict: the damage caused to roads and trails in forest and field by trotting horses, which on occasion brings complaints from others looking to recreate in the outdoors.

Attention must be paid to the following situations where conflict can arise between equestrian sport and the need of the environment for protection:

- When **choosing a site for an equestrian park:** Ecologically valuable habitats can suffer harm or even complete destruction from buildings and intensively used grounds (which entail draining and filling land, dividing cohesive natural areas, paving and covering surfaces, and placing stress on bodies of surface water and groundwater). The same holds true when equestrian parks and facilities are placed near nature reserves and other sensitive, ecologically valuable areas.
- When pasturage **is used too intensely**, for example, when less than a half hectare of pasture is available per horse.
- When **laying out riding trails** in ecologically valuable and sensitive zones such as wetlands, dry grasslands, coastal dunes, tidemarks, and ecologically sensitive forestlands.
- When **siting equestrian shows** and staging them: laying out routes for endurance competitions through ecologically sensitive areas, destroying valuable vegetation, leaving trash behind.
- When riders either intentionally or accidentally **leave marked trails** causing erosion on slopes, banks, and shorelines; damage to ecologically valuable vegetation from trampling.
- When riders and spectators **travel by car to and from** an event: noise and emissions from automobile traffic.

3. Finding Solutions

Outdoor, recreational horseback riding can be one of the environmentally friendliest forms of sport when the relatively minor ecological stresses outlined above are avoided.

Riding parks and stabling

Buildings (such as indoor riding halls, often featuring a restaurant or pub, horse stables, barns, and staff living quarters) can be sited in the landscape and operated in a way that takes the needs of the environment into account. Converting unused farm buildings into riding facilities is to be preferred over constructing new buildings. Ecologically valuable areas in a close to natural condition should never serve as sites for riding parks or for park buildings. In selecting a site for a park and designing its buildings care must be taken to ensure that the park and its structures blend in harmoniously with the surrounding landscape.

In **building operations** make use of state-of-the-art technology and the most up-to-date knowledge available in order to

- save heat (for example, through insulation),
- save electricity (for example, through appropriate placement of windows),
- make use of solar energy,
- reduce the use of drinking water, for example, through rainwater collection tanks,
- use environmentally-friendly cleaning agents,
- avoid generating trash and, where this is not possible, separate garbage and send all recyclable materials to recycling facilities.

The **design of buildings** should make use of all opportunities to green roofs and façades. Simple structures (such as nesting boxes) that would provide shelter for birds (such as barn owls, tawny owls, redstarts), bats and similar creatures dwelling and nesting in hollows, should also grace buildings.

A **landscape plan for the riding facility** featuring rows of trees and shrubs, hedges, copses, open meadows and planted embankments can serve both aesthetic and ecological purposes. Attention should be paid to the following principles:

- Instead of many small green spaces, use a few large green areas that are both pleasing to the eye and function ecologically.
- Instead of isolated green spaces, use a system of interconnected spaces (free of asphalt paths or roads) that link a number of biotopes with one another.

"Surplus Areas"

Those parts of the equestrian park that are not taken up by buildings or facilities offer significant opportunities to take environmentally sensible action:

- Instead of decorative lawns requiring a great deal of upkeep and maintenance, use large wildflower meadows maintained without herbicides that provide habitat for insects and other animal species.
- Instead of allowing facilities (such as pathways) to border directly on neighboring forests and bodies of water, leave semi-natural transition zones between facilities and surrounding natural areas (keeping a good distance between facilities and nearby forests, riverbanks or shorelines).

Keeping Horses

Feed for horses consists principally of oats (or other high-nutrient feed), hay and fresh grass from grazing. Pastures are given only a minimum of fertilizer because the grass should not have a high protein content. Mowing usually takes place when the grasses bloom in order to obtain the maximum amount of fibrous material (pastures are not ordinarily mowed a second time). Grassland species whose former ranges have been much reduced (for example, warblers, various kinds of butterflies, and a variety of meadow plants) find valuable habitat in extensive-use horse pastures, as such land is not treated with nitrogen or herbicides, nor artificially drained.

Not only can horses be raised in a species-appropriate manner on such land, they can also be used to keep grasslands open. Horses' selective grazing habits, combined with the minimal use of nitrogen fertilizers in extensive-use pastures, allow an ecologically valuable mosaic of various grassland plant communities to develop there, many of which are seldom seen today. Practices such as mowing and dragging, as well as periodically pasturing cattle on such land, provide for a sustainable use of the turf.

The following rule for **pasturing horses** holds true from an environmental point of view: the fewer horses per unit of pasture, the better for the environment. When low-intensity grazing takes place over a large surface area, rich and varied plant communities can develop (as well as the small animal communities that find their habitat in grassland). Such low-intensity use practices also allow the fencing off of valuable areas (such as areas of coppice and marsh) to protect their vegetation from being trampled or eaten.

A horse pasture should provide at least one hectare of grazing land for each horse in order to avoid undue stress on the turf (with resulting uneven growth and bald patches from trampling and overgrazing). Low-intensity use pastures, those providing at least two hectares of land per horse, make a positive contribution toward the ecological sustainability of grazing lands.

The practice of enclosing horse pastures with hedges should be increased. Even a hedge consisting of a single row (planted with hawthorn, red- and blackthorn) increases the ecological value of the pasture and provides aesthetic beauty. So-called "Benjes hedges" are even better. They consist of a 4 to 5 m wide and approximately 1.5 m high barrier thicket (an earthen embankment with shrubbery of various heights) with a single row of trees growing in the middle (Berger/Guba 1994). Such natural-like hedges are inexpensive to plant and offer ideal habitats for many animal and plant species.

Equestrian Events in the Open Country

Ecological problems can arise when racing routes head away from trails and agricultural fields and run through areas of ecologically valuable vegetation (such as heaths, dunes, or dry grasslands) or through the few remaining habitats of rare and shy animal species (such as grouse and birds that nest in meadows). Such conflicts can be avoided, however, when event organizers meet and consult with nature conservation authorities and local environmentalist groups while planning the route for a race.

The same holds true for steeplechases (or foxhunts), where riders gallop en masse along a pre-set course strewn with low barriers. These run in part through agricultural land, woods, heaths, and fallow lands. Marshy or boggy soils are avoided in all forms of cross-country riding for sports reasons. Riding through other types of near-natural landscapes, however, leads to serious conflicts with the requirements of nature conservation. These conflicts, too, can be nipped in the bud when the route for a steeplechase or foxhunt is set in consultation with the conservation authorities responsible for overseeing the area.

Crowds of spectators at large equestrian events can also cause problems for the environment. The careful siting of footpaths to keep people out of areas sensitive to trampling, proper disposal of trash, and sufficient parking located in non-problematic areas can ensure that noise and stress on the environment are kept within reasonable bounds (see chapter C).

GOLF (L)

1. Significance for the Environment

While golf is a popular and traditional sport in some countries (the USA, Canada, Japan, Scotland, where on the average between 6 and 10% of the population plays golf[15]), in others only a small number of citizens engage in it (for example, Germany, where in 1995 only 0.31% played golf).

15 In Ireland and Scotland the number of golfers in clubs or organizations comprise only 3.7 and 3.2%, respectively, of the population (Billion 1992).

Spaces and Facilities Used for Playing and Practicing Golf

Component	Use	Required Space
Fairway	Area between tee and green.	ca. 0.5 - 3 ha
Tees	Players hit balls from tees onto greens. Each fairway has 2-3 tees (men's, women's, and championship tees)	ca. 200 m2
Green	Area of the fairway specially prepared for putting balls and marked with a flag. 50% of all golf ball strikes occur on greens.	ca. 300 - 1,200 m2
Collar	Transition area between green and fairway. Players should be able to easily hit the ball from the collar onto the green.	ca. 300 - 1,200 m2
Roughs (semi-roughs, Roughs, hard roughs)	Roughs are supposed to contain playing areas and act as transition zones to neighboring playing areas. To make games more challenging, roughs can also be made to intrude into fairways. Hard roughs, located ca. 10-15 m away from fairways, have no effect on game.	Depends on overall size of golf course.
Hazards	Artificial hazards, such as sand traps, are distinguished from natural ones, such as grass hollows, ditches, and bodies of water. The type and number of hazards vary from golf course to golf course. For example, outside of coastal regions and heath lands sand traps can be reconfigured as putting traps or replaced by grass hollows in the fairway.	Depends on the type and number of hazards used. Sand traps, for example, can cover an area of between 5 and 500 m^2.
Driving range	Area for practicing hitting balls over a long distance.	2.5 - 4 ha
Golf academy	As in driving range	4 - 6 ha
Pitching area	Space for practicing hitting the ball over short and medium distances.	0.5 - 1 ha
Putting green	Space for practicing tapping the ball over very short distances (putting)	0.1 - 0.2 ha
Pitching and putting area with 6 fairways 9 fairways	Small golf course with short fairways, namely 50-120 m long. Here players practice hitting the ball to within a short distance of the hole.	3 - 4 ha 4 - 5 ha

From: Merkblatt "Naturschutz und Golfsport", LFU Bayern 1989

This chapter will only discuss golf courses as they affect the environment. It will not examine whether a particular golf course is suitable from the point of view of society – a very important question in some cases. This question arises when a considerable part of the land suitable for recreation within a given area (particularly when this land is located near urban areas) is given over to golfing. These courses then serve only the recreational needs of the relatively small numbers of golfers in the population.

Golf courses[16] usually have the following **forms of infrastructure:**

- A club house (with restaurant, management offices, lavatories, showers and changing rooms, golf shop)
- Sheds for storing equipment and parking maintenance vehicles (used by groundskeepers)
- Small shelters at teeing areas or for protection from rain
- Sheds for housing irrigation pumps
- A parking lot with 100-150 spaces (varying according to club membership)
- Entry and exit roadways
- Miscellaneous objects (such as markers on teeing grounds, flagsticks, teeing mats, benches, garbage cans, warning signs).

From an environmental standpoint, the **large amount of land** required to build a golf course represents a risk (in areas in a close to natural condition), but also a positive opportunity (in intensely utilized areas). The siting of a golf course will determine whether it benefits or harms the environment.

16 Normally, courses have 9 or 18 holes. 27-hole courses are rare.

2. Potential Conflicts

Checklist of Environmental Stress

The list that follows points out the potential impacts a golf course can have on the environment, arranged according to the source of the problem:

- Construction (clubhouse, ancillary buildings, roads, parking lot)
  - Covering the ground and vegetation with buildings or pavement
  - Dividing once contiguous habitats
  - Marring the visual appearance of the landscape
- Traffic (from golfers and visitors driving cars to and from the facility)
  - Noise
  - Exhaust emissions

- Altering the appearance of the terrain and altering vegetation cover (through use of fill dirt, excavations, changing existing vegetation)
  - Leveling terrain (on fairways, greens, and tees)
  - Piling up fill dirt on greens to create hazards
  - Filling and draining marshy ground and areas subject to flooding
  - Reconfiguring existing bodies of water
  - Replacing natural vegetation cover with lawns (for fairways, semi-roughs, greens, and tees)
  - Removing (clearing) trees and shrubs on the areas used directly for golf

- Disposing of sewage (from the clubhouse) and fertilizing greens and tees
  - Damage to surface bodies of water from channeling waste water into them or from runoff containing fertilizer
  - Contamination of groundwater from waste water or nitrate deposits (from intensively fertilized areas) seeping into the ground

- Maintenance and upkeep (of fairways, greens, tees, and to a certain extent, roughs); Destroying or placing stress on habitats by introducing changes
  - in the level of nutrients (fertilizing),
  - in the level of moisture (irrigating, draining),
  - through frequent mowing,
  - through the application of chemicals (destroying selected species through pesticides),

- in the porosity of the soil (compaction of the soil by heavy machinery),
- from wasting drinking water on irrigation, above all in drought-prone areas or on soils that have limited absorption capacity.

- Golfing activities
  - Disturbing animals requiring a great safety distance around them
  - Disturbing and placing stress on animals and plants by entering natural areas (ecological rest zones) adjacent to fairways

- Changing the appearance of landscapes having a special character (that should be preserved in their present state)

- Secondary effects
  - Slating the area around the golf course for the construction of vacation or second homes (a golf course would increase the attractiveness of a location such as an already densely populated Alpine valley).

3. Finding Solutions

a) Recommendations on selecting a site

The decision of where to site a golf course entails far-reaching effects on the ecology and aesthetic quality of the affected landscape.

A monotonous, "tidy" landscape consisting mainly of farm fields and other intensively used agricultural land has less ecological value than a golf course. The placement of a golf course in such an area can enrich it and more or less increase its ecological and aesthetic value. Thus, whenever possible, only intensively used agricultural land that is up for sale or lease should be considered for conversion into golf courses.

Sites having a diverse biotope structure are particularly problematic. Drought-prone regions should be avoided because of the great amount of irrigation water a golf course would require there.

The general rule for selecting a site for a golf course: avoid ecologically valuable areas, improve monotonous, intensively cultivated landscapes.

Areas with a high groundwater level or having porous soils with a low buffering capacity (such as sandy soils) are not suited for golf courses. The normal practices of lawn upkeep (applications of fertilizer and other substances) would endanger the groundwater in such areas.

When determining the suitability of a site for a golf course the greatest attention must be paid to the quality of its biotopes and to avoiding areas with sensitive soils. Other characteristics of the site that are relevant to the environment, however, also play a role, namely:

- Can the site be connected to the public sewer system (which would avoid the problem posed by insufficient treatment of waste water)?
- Can existing buildings be used (which would make the construction of new buildings unnecessary)?
- Is road access to the site already available (which would obviate the need for construction of new roads)?
- Could public transportation (trains and busses) easily serve the site?

The equally important criticism made against golf courses, namely that they claim open space for recreation that is already in short supply in and near urban areas, will not be discussed here, for it concerns social interests and not those of the physical environment.

b) Size recommendations for a golf course

The more land allotted to a golf course, the greater the chance

- of avoiding potential conflicts, since golfing activity can be kept away from ecologically sensitive areas (by making sure that sensitive areas lie away from fairways, or between fairways, and have sufficient buffer zones),
- of increasing the ecological value of the site: intensively-used land can be given over to extensive use or be withdrawn altogether from use; new habitats can be created (such as near-natural ponds and shrubby thickets).

Recommended Overall Size for a Golf Course in Relation to Length and Width of Fairways

	Average total length of fairways (km)		Total area of fairways (ha)		Total area of golf course (ha)	
Width of fairway including semi-rough	18 holes	9 holes	18 holes	9 holes	18 holes	9 holes
40 m	5,0	2,5	20,0	10,0	60,0	30,0
	6,1	3,1	24,4	12,2	73,2	36,6
50m	5,0	2,5	25,0	12,5	75,0	37,5
	6,1	3,1	30,5	15,3	91,5	45,9

c) Recommendations for laying out a golf course

Groups of shrubs (composed of species appropriate to the site) and bodies of water can run into the fairways and so function as natural "hazards." In addition, areas 60 to 100 m wide around tees may easily be allowed to grow as a meadow (a rough) receiving only a minimum of care. Such biotopes have admittedly only limited ecological value, as they are subject to a relatively high level of disturbance from nearby golfers. Yet they can offer relatively good habitat for species of animals and plants able to tolerate outside disturbance, and have a harmonizing visual effect on the surrounding landscape as well.

Golf courses should not be designed in a monotonous way, bat rather offer diverse habitats to animals and plants.

The width, shape, and sequence of fairways can be easily tailored to meet the demands and circumstances of the local environment. Moreover, "doglegs" – crooked or winding fairways – make it possible to avoid natural barriers (such as sensitive biotopes).

The boundary separating fairways and "ecological rest areas," (undisturbed biotopes) largely given over to nature, should not be a sharp one. Rather, golf course architects should make use of an opportunity to allow wide and diverse transition zones to grow between fairways and surrounding natural biotopes.

Sensitive areas of the golf course – those easily subject to disturbance – must be marked as such in order to keep golfers from entering them. Indeed, the rules of golf are entirely compatible with posting signs that regulate entry into these areas.

Further recommendations on design:

- Leave a sufficient buffer zone (at least 20 m wide), or rough, around so-called "tabu" areas requiring a high level of preservation. Woods should have unused fringes of land at least 10 m wide, covered with shrubs and perennials, along their edges.
- Do not dam streams in a near-natural condition.
- Restore streams that have been canalized (for example, made to flow through underground pipes or concrete channels) to their natural state. Close artificial drainage channels in former wetlands so that these may return to their natural state.
- Leave bodies of water with natural shorelines or banks alone, or create them and link them to surrounding habitats. The habitat of many amphibians consists of water, woodland and meadow together, which may not be cut off from each other by frequently mowed lawns. Bodies of water and adjacent zones should be connected with roughs ("biological corridors").
- Roughs composed of near-natural areas should not exist in isolated patches, but should rather run together to form continuous bands of close to natural land crossing fairways. Fairways should not cut through and divide swathes of near-natural land.
- Move as little earth as possible. Do not construct rectangular-shaped, raised tees (so-called "coffin lid tees").

- Leave sufficiently wide open spaces requiring little maintenance between fairways (such as sweeps of trees and shrubbery having sufficiently wide transition zones at their edges, strips of meadow that are mowed only twice a year, a natural succession of vegetation allowed to develop without interference).
- Do not install man-made structures (those foreign to the landscape) on the land such as fences and nets to catch balls.
- In protected watersheds and where groundwater lies beneath permeable soil layers: Lay a subterranean barrier beneath greens and tees, collect the seepage (in cisterns) and (by means of pumps) use it for irrigating these same areas.
- Do not channel water from tees and greens through drainage pipes into nearby streams or lakes.
- Use only species of trees and shrubs native to the site when laying out new plantings. Sow buffer and transition zones only with seed appropriate to the site.

d) Recommendations for buildings and infrastructure

- Use already existing buildings for the clubhouse and ancillary buildings. Incorporate any required new buildings into existing structures.
- Connect the clubhouse to the public sewer system.
- Roads should not cut through and divide areas of valuable habitat, nor should they cause damage to the edges of such areas. (Attention must also be paid to the effects roadbeds have on drainage and damming water flow.)
- Do not pave surfaces for entry and exit roads (use only water-permeable materials).
- Locate parking spaces sufficiently away from residences and recreational areas where peace and quiet is sought.
- The golf course should be served by public transportation systems.

e) Recommendations on Care and Maintenance

A **maintenance plan** should be formulated according to ecological principles that will stipulate how much and what kind of care the various areas of the golf course are to receive. The plan should be formulated in conjunction with the nature conservation authorities responsible for the area. Golf course developers should aim to have a plan that commits them legally to the principles of low-intensity maintenance ready for presentation during the official licensing process.

This will preclude certain conflicts from arising later. After construction, authorities should check to see that the plan is being carried out.

The following **developmental goals** for the care of roughs are desirable from an ecological point of view:
- A varied mixture of plant species: For the benefit of the animal world, aim at providing a great number of herbs and tall perennials, such that their periods of bloom cover the whole growing season. This will provide a nutrient base for butterflies and other insects that visit flowers, which in turn serve as food for birds, amphibians, and other animals.
- Structural diversity: vegetation of differing heights and densities creating a mosaic effect within a small area; variable small-scale terrain (for example, sinks or depressions in the ground); both these measures benefit creatures such as grasshoppers and mollusks.

Apply **fertilizers** only in the absolutely necessary amount. Fairways can do entirely without fertilizer approximately 2 to 3 years after they have been laid out. When the mown grass is left on the lawn as mulch, the nutrient level will remain in balance. It is worth noting here that on the average each hectare receives about 30 kg of nitrogen a year from the atmosphere.

As a matter of course **pesticides** should not be used. Their use should only be considered in exceptional cases and then for only a limited amount of time (for example, to treat an attack of mold on greens). We recommend instead biological and mechanical means of pest and weed control, such as providing nesting sites for birds, and eliminating weeds through mechanical and thermal methods.

Irrigation should be limited to the absolute minimum in order to conserve water. When possible, water from rainwater collection tanks should be used instead of drinking water.

The **groundskeeper** should have sufficient formal education in ecology. He should keep a log recording treatment and maintenance (including any applications of fertilizer) given to the grounds, and most particularly any applications of pesticides (a "poison logbook").

MOTORSPORTS (M)

1. Significance for the Environment

The following motorsports contests and races are held on various kinds of courses and terrain:

Automotive sports

Track races, slaloms, go-cart races, rallies, truck races, and skill competitions are held on special surfaces such as racing tracks, very large parking lots, or paved circuit courses. Various other kinds of races, such as those leading through mountainous terrain, orienteering races, and rallies through the countryside all use public roads and highways. Autocross, auto trials and (occasionally) rallies are held on open terrain: on gravel beds, sand, and harvested cropland, in gravel pits, quarries, on fallow agricultural land and in forests.

Motorcycle sports

Track races and, occasionally, street races and reliability runs, are held in special facilities such as race tracks and stadiums, and in indoor racing halls. Road racing, motorcycle competitions, slaloms, and various other kinds of motorcycle races and tests (such as those in which contestants set out from various starting points, rallies held in the countryside, and road tests) use public roadways and parking lots. Trial and endurance competitions, motocross, and cross-country racing all take place in the outdoors (in gravel pits and quarries, on fallow and harvested fields, and on dirt farm roads).

Aside from their harmful effects on the environment (see chapter 2 on this matter), motorsports are fraught with controversy for environmentalists because they encourage the use of motorized forms of transportation (quite apart from racing purposes), and so indirectly contribute to the high level of stress that recreation-related driving puts on the environment.

2. Potential Conflicts

The main conflicts between motorsports and the requirements of nature and the environment for protection fall into two categories: pollution of the physical environment (either through noise or exhaust emissions) or the degradation of ecosystems (from driving off of paved roads in ecologically sensitive areas).

The North German Technical Monitoring Association (Technischer Überwachungsverein Norddeutschland) and other institutions (cited in UBA 1989) have recorded emission values for noise (sound intensities) at motorsports facilities as follows:

Motocross: 128 dB(A) when motorcycles are in operation for 70% of the total time of a sporting event.

Speedway racing: 135-138 dB(A) when motorcycles are in operation during 40% of an event.

Sand track racing: 137 dB(A) when motorcycles are in operation during 40% of an event.

The following three kinds of motorsports make intensive use of open countryside:

- Cross-country races with endurance bikes, which take place on paved roads, unpaved farm and forest roads, and (in the case of cross-country test runs) off-road on various kinds of terrain.
- Trials on difficult terrain, which test only contestants' skill and agility.
- Motocross, in which contestants speed en masse along a circular route without any extremities in terrain (only bumps in the ground and other similar light obstacles are needed).

Use of near-natural areas (such as forest, fallow agricultural land, abandoned rock quarries, and low-intensity use pastureland) for motorsports produces serious ecological stress such as destruction of vegetation, compaction of the soil, the running over of valuable animals or their displacement from the affected area.
Not only sporting events can cause serious damage to sensitive terrain, but also training rides and, increasingly, the independent recreational activities of motorcycle owners.

Motorcycle clubs and organizations can work to stop undisciplined, environmentally reckless off-road driving by urging their members to practice motorsports only on surfaces officially designated for them.

3. Finding Solutions

- Additional measures for vehicles (beyond those already in force)
  - Reduce fuel consumption (saving energy).
  - Reduce pollutants (by requiring catalytic converters, using unleaded gasoline).
  - Reduce noise by designing vehicles and components that run more quietly.
  - Manufacture vehicles and parts that last longer (thereby reducing the amount of waste)
  - Service vehicles regularly.
  - Use products or parts made from recycled materials.

  In the following section we shall consider how to choose an event site and plan a race route to avoid harming the environment.

- Ecologically suitable sites for cross-country races

  Which kinds of areas are suitable from an ecological point of view for cross-country motorsports? Proposed areas have to satisfy two conditions. First, the area may not have any ecologically valuable features or qualities that motorsports activity would harm. Second, the area has to have the kind of terrain that would offer the necessary challenge and degree of difficulty required by cross-country auto or motorcycle racing.

The following areas satisfy both of these conditions:

- Gravel pits, sand pits, and rock quarries either still in use or recently taken out of operation. Artificial barriers and hazards can be set up as necessary.
- Open dumps, landfills for non-toxic waste as well as derelict industrial areas. They often already contain areas suitable for motorsports or offer good possibilities for creating such areas.
- Military bases that for certain periods of time open their gates to motorsports enthusiasts. Motorsports may then take place in areas of the bases ordinarily used for testing military equipment and vehicles.
- Sports halls and stadiums. These buildings could provide a space for cross-country competitions more often than they have in the past.
- Man-made practicing grounds.

As a matter of course all parts of the landscape in a close to natural state (such as forest, fallow land, marshy, and dry areas) should be closed to motorsports. Proper maintenance of vehicles is essential, for it will prevent oil from leaking out onto open land and thereby posing a danger to groundwater.

- Conserving resources and appropriate behavior
  What can the person who engages in motorsports do?
  - Never drive in areas not specifically designated for motorsports.
  - Never dump trash, oil or gasoline out in nature (a punishable offence!).
  - Do not take part in events lacking proper permits.
  - Reduce emissions of pollutants.
  - Reduce the use of gasoline and, as far as possible, use lead-free fuel and low-emissions vehicles.
  - Have the highest degree of consideration for biotopes and the needs of nature.
  - Use environmentally sound motorsports vehicles and equipment.
  - Take part only in appropriately organized events having the necessary permits.
  - Avoid unnecessary test runs as well as letting engines idle needlessly.
  - Only optimally tuned vehicles at the starting line: "souped up" vehicles harm not only the environment, but give motorsports a bad reputation.
  - Do not clean vehicles outside in nature and clean them only with biodegradable substances.
  - Dispose of used motor oil, coolant, and related fluids separately from other waste.
  - Work on engines and transmissions only on asphalt or concrete surfaces or with protective covering on the ground.
  - Clean and remove all forms of soil contamination resulting from accidents and spillage.
  - Avoid the following: producing unnecessary noise that would annoy neighbors; damaging soil, plants, and the environment; causing ruts or troughs in the ground; leaving behind trash; and contaminating soil and water. Point out inappropriate behavior to others.

The National Commission for **Automobile Sports** in Germany (ONS) has issued compulsory rules governing noise and exhaust emissions. As of January 1, 1997, they are binding on all automobile-racing events approved by the commission and are supposed to reduce the amount of noise and emissions produced at races (ONS 1997).

The International Motorcycle Sports Commission has published a comprehensive "environmental code" that provides regulations and recommendations for **motorcycle sports.** The code's goal is "to positively shape the interaction between motorcycling and the demands of the environment." The regulations and recommendations address in particular

- problems involving noise, fuel, the protection of soil, and the cleaning of vehicles and equipment,
- the behavior of spectators, event organizers, circuit and track managers, and motorcyclists sharing the roadways with other motorists.

The code has an appendix that contains checklists for various motorsport disciplines as well as for motorcycle tourism (OMK 1997).

AERIAL SPORTS (N)

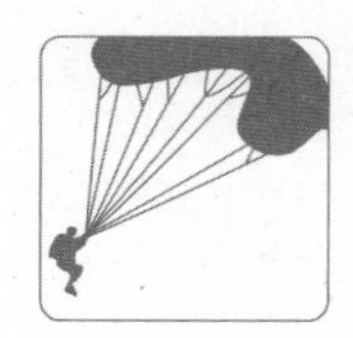

1. Significance for the Environment

Here we examine sports and recreation involving airplanes, gliders, ultra-light planes, hang gliders, parachutes, balloons, and model airplanes.

Facilities for aerial sports (excluding fields for model airplane flying) are found only in the open countryside. Spaces for flying airplanes are located outside of cities and towns in order to shield residents from noise. Climatic features (areas having thermal winds) and accessibility (proximity to urban areas) largely determine the locations of facilities for non-motorized flying.

Both conflicts and ways to solve them depend to a great extent on the specific aerial sport involved. The following points do not address any specific discipline within the broad category of aerial sports. They look instead at general areas of conflict that to a greater or lesser degree arise from the use of various aerial craft.

2. Potential Conflicts

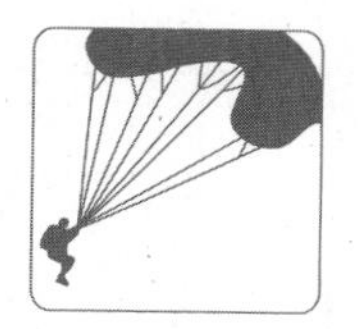

When too little heed is paid to the demands of nature and the environment, the following problems can be expected:

- Problems arising from the construction and operation of facilities for aerial sports:
  - Selecting a site without paying sufficient attention to the surrounding environment can place stress on, or even destroy, relatively untouched animal and plant habitats.
  - Maintenance practices that are far removed from natural cycles can damage areas of vegetation subject to little use (for example, mowing meadows too frequently or at improper intervals).
  - Spectators at large aerial events or shows can damage vegetation and the top soil layer through walking as well as by driving and parking their cars.
  - Buildings and other structures can mar the natural scenery of an area.

- Stress on humans (noise) and on animals (disturbance) from flying:
  The following aspects of flying can negatively affect animals:
  - Sound that raises the overall level of background noise can interfere with animals' ability to communicate with one another, to find food or to sense approaching predators.
  - Flight movements can distract animals, disturb them or cause them to take flight, and interrupt activities such as breeding, grooming, and searching for food.
  - Human activity at starting and landing fields can damage the vegetation in these areas (trampling).
  - People and activities associated with the actual flying (such as spectators, assistants, people arriving and departing in cars) can generate noise and trash and trample vegetation.

  The noise and pollution produced by motorized aircraft can affect humans.

Disturbance and Displacement of Animals

Aircraft can have a disruptive effect on certain species of animals when it does not keep a sufficient distance away from animals. This happens, for example, in taking off and landing: planes then fly low over animals living near the airfield.

The protected animals most likely to suffer disturbance from aerial sports are birds that nest in (humid) meadows (such as meadow pipits, curlews, black-tailed godwits, snipes), birds nesting in rocky areas (such as alpine wall creepers, alpine choughs, peregrine falcons), grouse (black grouse, capercaillies), golden eagles, eagle owls, and birds nesting in treetops such as ospreys and hobbies.

Aircraft can partially or completely disrupt such animals' success in breeding and reproduction as well as limit their ability to find food. The relatively few hours in the day in which flights take off and land are enough to cause such negative effects (for example, by allowing eggs to grow cold).

Whether a potential conflict actually becomes a real one or not depends on whether, and under which conditions, aircraft have a more or less permanent **disruptive effect on animals.** The ecological factors making such disruption possible will be briefly described below in order to show the means by which a proposed airfield or a given activity may be gauged for its "compatibility with nature."

Many vertebrates serve as potential prey to aerial predators, and are internally "programmed" to know how to escape from them. A regularly hunted species must know how to quickly identify potential attackers and how to differentiate dangerous aerial predators from harmless flying objects (such as various forms of recreational aircraft).

To start, we must differentiate an innate from a learned ability to recognize predators. Certain reactions to aerial predators are genetically determined. Here an "innate trigger mechanism" is at work. In such a case an animal's reaction to a given stimulus takes place automatically, and thus does not have to be learned. Animals that are potential prey must be able to react immediately and appropriately to the sudden appearance of a predator if they are to survive. No time remains then for them to more precisely identify a flying object. Thus the innate trigger mechanism responds rather unselectively to simple stimuli.

That some species take to flight when harmless objects (such as kites, hang gliders, model airplanes) appear overhead results from the animals mistaking them for aerial predators. The instinct to flee is activated when a flying object (a "predator") travelling at a certain speed above the ground, and located a certain distance away (the safety distance), enters the animal's field of vision.

Most of the time the experiences gained by an individual animal over the course of its development affect and complete this innate trigger mechanism. An animal that is potential prey can fine-tune its perception and responses through such learning experiences. The innate trigger mechanism, which at first caused the animal to react without selectivity, becomes through experience a "modified trigger mechanism" much better able to differentiate stimuli and react selectively. Thus learning is above all the reaction to harmless species or aircraft. An animal's initial movement towards flight is suppressed when it has already learned that the identified object poses no danger: familiarity now enters the picture.

"Harmless" flying objects can also cause animals to take instantly to flight when the object is flying so low that the animals lack sufficient time to identify the object as harmless,[17] or when they lack the experience that an object resembling a predator is actually harmless. Whether an animal takes flight or not also depends on whether the terrain offers sufficient cover. The more often animals meet with a given flying object and so become familiar with it, the less time they need to identify it.

The sighting of an aerial predator (whether real or supposed) constitutes a disturbance when the animal interrupts its normal activities during this time (such as looking for food). This also holds true in situations where the animal does not take flight but reacts more subtly (experiencing, for example, increased heart rate or anxiety).

The level of disturbance also depends on the particular flying object, that is, on its size, on the noise it generates and on its distance from the animal. The frequency of flights (their regularity) and the degree to which animals can foresee aircraft's movements (change in direction) also play a large role here. On the other hand, the shyness of a particular animal (or the degree to which it has become used to the disturbance) and the terrain of an area also affect its response.

---

17 Animals take to flight in especially great panic when they first notice a "predator" after it has already intruded into their perceived safety sphere (for example, in terrain lacking natural vantage points). A low-flying "predator" also triggers such a response.

3. Finding Solutions

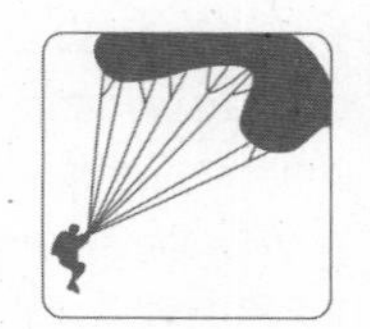

Minimizing conflict through use of space

When direct flying over a given animal habitat is unavoidable, planes and other craft must maintain a minimum height over the area that corresponds to the safety sphere animals need around themselves. This safety distance[18] measures between 150 and 200 m. The minimum distance to observe when approaching flocks of birds increases by 3 to 5 times, particularly when these congregate on open water surfaces lacking vegetative cover (such as shoreline reed beds).

Takeoffs and landings must always follow the same, specified routes so that the affected animals can reckon with them and grow accustomed to them (determining exact takeoff and landing routes). Pilots must be informed accordingly.

The following measures can help prevent conflicts between those engaging in aerial sports and wildlife that is particularly susceptible to disturbance:

Selecting a site for takeoffs: Places for takeoffs should not be located where flight movements can be expected to disturb animals in the vicinity that need protection.

Proper flying: Appropriate routes for takeoff and further flight can keep aircraft sufficiently away from sensitive habitats and ensure that they do not intrude into animals' safety spheres by flying too low. Air sports enthusiasts, moreover, must be given information on the reasons for these measures. Their cooperation should also be requested.

---

18 The safety distance is the minimum distance an animal requires between itself and a potential intruder to remain undisturbed. It is, for example, the distance that a breeding bird needs around itself in order to go to its nest and remain sitting there, or that a chamois needs to not run away when a flying object passes over it (the fly-over altitude).

Minimizing conflict through scheduling

Flight activity that remains relatively constant and regular over the course of a year is less problematic to wildlife than activity that begins suddenly and continues only over a short period, for example in spring or as a part of sporting events.

Limiting flight activity to between 8:00 a.m. and 6:00 p.m. greatly helps to reduce disturbance of wildlife to a bearable level. Animals then grow accustomed to having the same period every day free of disturbance and can make use of this time.

Nesting birds are most sensitive in spring and early summer when they are breeding and raising their chicks. Consequently, greater distances must be kept from animal habitats during these times.

The demand to consolidate airfields that is sometimes heard does not make sense from an ecological point of view. Such centralization brings increased use of the facility on days suitable for flying, gliding, and other air sports.

Making sports airfields more ecologically valuable

Without endangering flight safety, the ecological quality of sports airfields can be raised by restoring degraded areas to a more natural condition. While keeping the technical requirements of flight safety in mind (for example, preventing birds from striking aircraft), we can develop the different parts of the airfield to harmonize with nature.

- Areas can be turned over entirely to nature (natural succession), or
- biotope management can be used (employing only methods of upkeep that accord with ecological principles).

Constructing such habitats allows valuable animal and plant communities to develop and establish themselves at the airfield site. The safety areas and edges of airfields have a certain "ecological advantage" in that they are not used for agriculture and people rarely set foot on them. Thus these areas constitute an ideal refuge for species of (small) animals that are not disturbed by flight activities.

Protecting and conserving natural resources

Air sports enthusiasts can do this by

- properly disposing of waste such as used motor oil, and by separating trash,
- doing everything possible to further the development of motors that generate less noise, emit less pollution, and conserve energy.

SNORKELING AND SCUBA DIVING (O)

1. Significance for the Environment

There are basically two kinds of skin diving:

a) Snorkeling with basic gear: swimming just below the surface of the water while occasionally diving deeper, using a mask, snorkel and flippers, for purposes of better underwater viewing.

b) Scuba diving: While underwater the diver breathes compressed air from a cylinder through a demand regulator. The diver can thus remain longer underwater and can observe aquatic life at greater depths. Scuba diving is generally regarded as a sport.

Skin diving (and its specialized form, the orienteering dive) takes place in both natural and artificially created bodies of water (natural lakes, reservoirs and excavation pits, rarely rivers). Training is usually in large indoor or outdoor swimming pools.

Competitions at various levels (state, national, European, world championships) are held in the disciplines of swimming with fins, underwater distance swimming, orienteering, and underwater rugby. Large indoor swimming pools are most often used for these contests (with the exception of the orienteering dive).

Skin diving belongs to those sports that are at times practiced in natural settings. In order to view underwater plants and animals, people seek waters high in clarity and offering a richness of species and ecological structures.

2. Potential Conflicts

Although the environmental stresses discussed below arise from skin diving alone, one should not view them in isolation from other demands placed upon bodies of water. Among activities that place stress on bodies of water, skin diving often has only a minor role, for example in the case of a lake frequently used for recreation which receives nutrient-rich runoff from surrounding farmland (eutrophication). Much more serious, however, is stress occurring in and around waters that are either principally or exclusively used for diving, and which are low in nutrients and provide a home to particularly sensitive plant and animal communities.

The following outline briefly lists types of the stress that skin diving can place on the environment:

- Stress arising from **driving** to a diving site and parking there (noise, exhaust emissions, resource consumption).

- Stress on **shoreline areas in a close to natural state:**
  - Trampling shoreline vegetation when camping, or entering and exiting the water.
  - Destruction of aquatic plants that grow in shoreline reed belts, in the floating plant zone, in the submersed vegetation zone, as well as in underwater meadows (such as stoneworts). Damage may occur in the form of trampling or tearing (when vegetation gets caught in diving gear) and may result from divers' movements in the water.
  - Disturbing waterfowl in their refuge areas as well as in their nesting and molting grounds by entering shoreline areas (during breeding season) and the open-water surface (during molting season).
  - Destruction of fish and amphibian spawning grounds in shallow water.
  - Causing noise when filling compressed air cylinders and blowing air through valves.
  - Leaving trash behind.

- Stresses **on bodies of water:**
  - Damage to the particularly rich plant and animal communities living in the shallow water zone (from stirring up sediment, trampling, and compressing the lake bottom), particularly where divers enter and exit the water.

- Negatively affecting photosynthesis as well as releasing nutrients through stirring up sediment by swimming with flippers near the lake bottom.
- Divers' exhaled air causing nutrient-rich water from lake depths to rise upward: transport of nutrients from the hypolimnion into the epilimnion. This effect occurs only in eutrophic and polytrophic bodies of water.
- Underwater hunting: danger of causing extinction to certain fish species and disturbing the balance between predators and prey (plays no role in Germany).
- Disturbing fish during their periods of rest (at night and in winter).
- Disappearance of microhabitats (niche animals).

Diving tourism in tropical and subtropical countries

The main destinations for divers from central and northern Europe on vacation are tropical coral reefs and the Mediterranean. The rich variety of underwater species here, the high underwater visibility, and steady sunshine constitute an enormous attraction that draws large crowds of divers making countless dives. Marine parks have been established to protect sensitive and precious underwater ecosystems from destruction, for example at Ras Muhamed on the Red Sea in Egypt, La Ravellata on the Spanish Mediterranean, and L'Estartit in Corsica.
In some areas (particularly in the Mediterranean) we can already see the catastrophic effects of harpooning ranging from the extinction of rare species (some species of sea bass, for example) to the creation of lifeless undersea "deserts."[19] Such practices also ruin recreational divers' experience of nature, for fish are especially shy in areas subject to a great deal of harpooning. Divers can observe fish here only with great difficulty, as fish in these areas keep an exceptionally great distance away from man.

Destruction and damage to coral reefs from diving constitutes a further problem. Fragile coral is damaged when people step on it, cast down anchors, and intentionally or accidentally break off coral pieces. Sediment cast up by divers' movements can hinder, or altogether prevent, the growth of coral.

19 Recreational divers rarely take part in underwater hunting: on most coasts divers are not permitted to harpoon. There are, however, still underwater hunting contests. Most hunters are local people who sell their catches to restaurants.

3. Finding Solutions

Three methods of problem solving are available here: 1) zoning, 2) using infrastructure to channel and direct visitors, and 3) providing education and establishing codes of conduct.

• Zoning

A foresighted use plan should identify waters that are both attractive to divers (whether throughout the year or only at certain times) and that from an ecological point of view can support diving activities. It should also clearly designate those bodies of water, or portions of them, that are ecologically valuable and susceptible to damage from diving and related activities (declaring them either "limited use areas" or "tabu zones").

From an ecological perspective, diving is **unsuitable** in the following bodies of water due to their extreme sensitivity and high ecological value:

- Small lakes with completely intact shoreline vegetation (reed belt, floating plant zone, and submersed plants) and without a place where people can directly enter the water, such as a dock
- Small lakes with valuable animal and plant communities needing protection
- Protected bodies of water where diving would conflict with the goals of the protective measures in place.

The above bodies of water are, from an ecological perspective, also unsuited for a variety of other recreational uses (such as sailing, windsurfing, swimming, fishing).

In other bodies of water diving should be prohibited in **certain parts** and at **certain times** when divers

- could disturb spawning fish or destroy spawn,
- could disturb an area of water to which molting waterfowl have retired,
- could negatively affect breeding waterfowl,
- would have to swim across underwater meadows (featuring stonewort or characins) at a minimal distance above the lake bottom.

• Visitor management

Prohibitions are by no means the only way to keep divers from practicing their sport in ecologically unsuitable bodies of water: **attractive alternative sites** for diving can be provided that will draw divers, especially when the sites offer appropriate infrastructure. On the other hand, a lack of necessary infrastructure (such as lack of road access to a diving site, road closures) can create obstacles that are difficult to overcome and so make easily accessible bodies of water even more attractive to divers.

Large diving events or competitions (swimming with flippers in open water and orienteering dives, for example) should only be held in bodies of water that can weather the event without suffering ecological damage. These waters may, for example, already be far removed from their original natural state or have no animals or plants requiring protection.

Visitor management on a small scale is also possible. Planners can direct diving activities into ecologically suitable areas and protect sensitive areas by designating and clearly marking entry and exit points into and out of the water as well as directions in which divers should swim. The same methods can also prevent people from making their own trails.

Ways to channel divers briefly described:

- Indicate bodies of water suitable for diving (such as lakes that have formed in excavation pits).
- Provide easy access (roads or trails leading to the site) or, where necessary, make access difficult.
- Designate areas for parking and for picnicking and prohibit parking elsewhere.
- Install toilets and trash containers so that they are easily accessible.
- Identify points of entry and exit into and out of the water (docks, artificial gravel banks) and make them known to the public.
- Plant thorny bushes to prevent people from making their own trails.

- **Education and codes of conduct**

To a large degree, appropriate behavior by sports enthusiasts in a natural environment depends on their knowledge about this environment and about how it will react to their intended sport activity. What makes a particular body of water ecologically valuable and sensitive to diving? What needs protection and what kinds of behavior endanger the protected qualities? As part of diving instruction courses and involvement in club activities, divers can be taught how to systematically observe bodies of water (and so to recognize and report problems) and how to care for and restore the landscape, for example,

- how to plant underwater plants after removing weeds,
- how to make trails to the diving site secure and stable,
- how to reduce existing unauthorized trails,
- how to restore eroded banks and shoreline areas,
- how to plant reed beds in the shallow water zone.

By observing the following rules of conduct divers can avoid placing stress on nature, or at least keep it to a tolerable level:

The Association of German Sports Divers (VDST) issued the following "Golden Rules" for divers:
"Sports divers

- don't walk or swim into reed beds,
- don't destroy shoreline vegetation,
- protect breeding and nesting areas,
- move smoothly through the water and don't stir mud up from the lake floor,
- don't damage underwater vegetation,
- use existing trails and docks,
- observe conservation regulations,
- protect spawning grounds and wildlife resting areas,
- keep cars and compressors away from the water,
- keep water and shorelines clean, because sports divers treat nature fairly."

CANOEING (P)

1. Significance for the Environment

Canoeing comprises the following sports disciplines:

- Racing: covering distances over calm water in the shortest possible time
- Marathons: at certain points along the racing route contestants must pick up and carry boats, otherwise as above
- Slaloms: gates mark the racing route over fast flowing water
- White-water racing: in rapidly flowing rivers with obstacles
- Canoe sailing: uses special canoes outfitted with sails
- Canoe polo: ball game on the water
- White-water rodeo: tracing acrobatic figures with the boat on moving water
- Canoe competitions that combine other sports with canoeing.

The following forms of canoeing are for the most part practiced outside of formal competitions and contests:

- Rafting: riding down white-water rivers in rubber boats, normally offered through commercial ventures.
- On certain stretches of rivers people may rent canoes for pleasure trips. They traverse an agreed-upon route for a day or half-day tour.
- Touring by canoe: The great majority (90%) of canoeists in and outside of canoeing clubs practices this form of canoeing. Participants in this activity traverse various rivers and lakes in kayaks or Canadian canoes. Most of the time people travel together either in small groups composed of family members or young people. Canoeists normally make such trips between March and October. To the extent that weather conditions permit, people travel in the other months as well with protective clothing. The number of canoeists during this time, however, is much lower than in the summer months.

Kayaks and Canadian canoes are normally used in canoeing. A kayak, usually ca. 4.50 m long, is enclosed all around with the exception of the cockpit opening where the canoeist sits and rows with a double-bladed paddle. A Canadian canoe is an open boat that is steered with a single-bladed paddle.

As a popular form of recreation touring by canoe is very much tied to nature. From an ecological perspective, however, canoeing is problematic (in contrast to rowing) inasmuch as canoeists can paddle along small, relatively untouched rivers. With the durable boats presently available people can paddle down streams including those with shallow gravelly beds without harm to the craft. A one-man kayak with a flat oval-shaped hull has a draught of only 5-10 cm (without baggage).

More durable materials (such as polyethylene) have also increased the arena of action for white-water rafting. People can now travel down strong, rushing rivers heavily punctuated with boulders, steps, falls and rapids without worry of damaging the boat.

2. The Ecological Significance and Value of Watercourses

Streams and rivers have great ecological significance for our landscape. Construction along watercourses and lakes and the discharge of substances into their ecosystem (including bottomlands and oxbow lakes) have caused relatively untouched, still natural riparian biotopes to become rare. Both flowing and still bodies of water in a close to natural condition constitute an important "backbone" of biological corridors, which, in turn, prevent various habitats occurring over a large area – both aquatic and non-aquatic ones – from becoming isolated from one other. Protecting and restoring our system of lakes and rivers constitutes one of the central tasks of nature conservation.

The following criteria can help us determine whether a given body of water requires protection and how sensitive it is to disturbance:

- Rarity of habitats (for example, stream banks and lakeshores free of construction, reed beds, bottomland forests, tide flats, and sea-inlets, all of which could be destroyed by human activities)

- Presence of endangered animal species (included in the Red List), for example
  - Species living in and around wild rivers: sandpipers, little ringed plovers, terns, goosanders, gray wagtails
  - Species occurring in and near streams of lower elevation mountain ranges: water ouzels, kingfishers, pearl mussels, and fishes and insects that live or lay their eggs in the gravelly or sandy substratum of streambeds
  - Animals living in and around lowland rivers: beavers, otters, black storks, sea eagles, species of dragonflies
  - Animals that live in and around lakes or on the coast: great reed-warblers, ducks while they raise ducklings and molt, wading birds, seals.

- Occurrence of plant species (or communities) along shores and banks or in the water (for example, reeds, and plants with floating leaves) that need protection.

3. Potential Conflicts

We can distinguish the following factors that can lead to stress on the environment:

- Damage to vegetation along shores and banks from trampling
- Compression of the soil from canoeists making trails as well as resting sites in river meadows
- Damage to water plants from paddling in shallow waters
- Endangerment of the spawn of fish, amphibians and small creatures living on the stream or lake bottom from paddling that stirs up sediments into the water
- Disturbance to animals on the water or shore (arising from canoeists simply being present in an area, or from attracting attention to themselves while coming too close to wildlife)
- Leaving trash and feces behind in resting spots
- Incursions into nature from canoeing-related land development such as widening of trails, construction of parking spaces, campgrounds and sanitary facilities.

These factors result in serious environmental stress only when they occur in ecologically sensitive and valuable landscapes and habitats (such as bottomlands, gravel banks with rare species and populations, valuable plant communities, smaller watercourses with animal species susceptible to disturbance living in the streambed).

We may distinguish environmental stress according to the activities that cause them: stress arising from driving to and from the canoeing site, that arising from hauling boats into and out of the water, and that stemming from paddling along the chosen course, including stopping and disembarking to take rests.

4. Finding Solutions

The following measures can prevent conflicts from occurring in or around lakes and watercourses worthy of protection:

Planning

Returning watercourses and lakes to their previous natural state (renaturalizing), and increasing their structural complexity, stand as the most important tasks for planners. Besides bringing desirable ecological and aesthetic effects, such planning methods can make less sensitive watercourses more attractive to canoeists. By offering them more such sites, canoeists can to some degree be steered away from sensitive bodies of water.

But as long as such large-scale programs of ecological improvement remain dreams and hopes for the future we must focus on formulating rules to govern canoeing in sensitive watercourses and lakes. The following elements can help, taken either alone or in combination, to formulate such rules of conduct. Here the particular situation and qualities of each body of water must also be kept in mind.

- Prohibiting canoeing in waters either too shallow or too narrow for it: Here **minimum values for the width and depth of waters open to canoeists** must be set. Natural watercourses (those that have not been canalized) should have a width of at least 3 m and depth of at least 20-40 cm, depending on the current, for them to be open to canoeing.
  The goal of this minimum depth standard is to decrease the chance that boats and paddles will come into contact with the streambed.
  People should also observe this standard of minimum depth when traversing the shallowest areas of watercourses having sufficient width. This standard does not apply, however, to streams that have been completely canalized, since these do not meet the criteria for environmental protection.

- Designating the **direction in which canoeists may paddle:** When canoeists attempt to steer away from oncoming boats, they may touch the streambed and thereby cause ecological damage in sensitive waters. This is particularly a problem on watercourses that are very popular with canoeists. To protect watercourses, paddling in an upstream direction should be prohibited in relatively narrow, swiftly flowing streams.

- **Indirect ways of channeling canoeists** to unproblematic areas: An intentional lack of **parking places and trails** can make access so difficult that many

canoeists are kept from using sensitive areas of watercourses (particularly the upper reaches of streams). Such a policy can work against crowding on watercourses when the central problem threatening the stream stems from its great popularity with water sports enthusiasts.
By the same token, ease of access (through the provision of trails, parking, rest areas, and boat launching sites) to bodies of water that are able to withstand the effects of canoeing can be used to attract canoeists to these unproblematic areas.

- **Closure** of watercourses to canoeing, **either at certain times or in certain parts:** In order to protect sensitive or rare species, it may be necessary to close a particular body of water either altogether or for certain times of the year or day (for example, during the breeding season of a particular bird species) to all kinds of water sport activity. Such closures must be carefully prepared and the scientific reasons for them explained to canoeists, for this measure directly and painfully limits canoeists' freedom of movement. Even closures of just small stretches of watercourses disrupt and reduce the quality of the entire area used for canoeing. Complete closures of watercourses are thus always a measure of last resort, to be employed only when weaker measures have not worked sufficiently to protect nature.

Canoeists' Conduct

The following rules of considerate sports activity aim to ensure that measures and policies established by planners to protect the valuable ecological qualities of certain bodies of water are followed, as well as to obviate more limiting and drastic measures. What exactly can canoe clubs and individual canoeists do to protect riparian and aquatic environments?

- When planning a trip canoeists can inform themselves about the streams or lakes they will use, about possible closures or regulations, about the water level, sites for launching boats into the water and hauling them out, and about rest areas.
- They can inform fellow canoeists about alternative streams in a given area that are not sensitive to canoeing.
- They can provide important information about ecology to other canoeists, as well as heed it themselves, and organize their activities accordingly (for example, by observing any seasonal or spatial restrictions).
- Do not drive cars on top of bank and shoreline areas in order to transport boats to the water (except where designated sites for this exist); do not park cars off roads in non-designated areas.

- Avoid sensitive shoreline vegetation such as reed beds and areas of plants with floating leaves.
- When paddling, keep the necessary distance from relatively untouched and intact shorelines and banks; only leave designated canoe routes when absolutely necessary.
- Keep away from areas where waterfowl are swimming with their young or breeding.
- Always make sure that the water level is high enough.
- Do not land boats on shoreline or bank areas in a natural, untouched state and use them as resting sites.
- Do not paddle where a great number of canoeists are already on the water.
- Do not go canoeing in large groups.
- Avoid the permanent presence of canoeists on the water; allow rest periods (this will prevent disturbances from stretching over a long period of time).
- Avoid canoeing repeatedly through certain stretches, especially when several groups of canoeists are on the river.
- Do not hold rallies and large canoeing events in sensitive waters.
- Take trash with you and dispose of it in an environmentally sound way later.
- Avoid causing unnecessary noise and commotion.
- Never pitch tents or set campfires outside of designated sites.
- Make note of areas of contaminated water (from illegal sewage pipes, for example) and report them later.

As a matter of course canoeing clubs should encourage their membership to act responsibly in nature by providing information about ecological cause and effect relationships, and the purposes and goals of nature conservation measures.

Every canoeing organization should take the necessary pains to protect the environment, even when this means limiting its activities voluntarily. This will make regulations and prohibitions on use, such as closures, unnecessary.

Commercial canoe rentals also have a responsibility here. They should inform sports enthusiasts about appropriate behavior and only list waters that are not ecologically problematic in information they give out to customers. Where waters have only a limited resilience to the negative impacts caused by canoeing, it may be necessary to limit the number of boats (with appropriate identification tags or plates) a canoe rental may operate on the water.

SAILING AND WINDSURFING (Q)

1. Significance for the Environment

Sailing and windsurfing differ from each other from an environmental point of view in the following ways:

- Sailboats normally require a marina, which (in contrast to launching sites for windsurfers) at times entail serious and massive alterations in shoreline areas.
- Because surfboards (in contrast to boats) have a minimal draught, they can be used in areas of shallow water. Thus in windsurfing, the sports activities themselves entail a greater risk of environmental damage than do the sports facilities windsurfers use.

Because both sports share many of the same points of contact with the field of natural and environmental protection, sailing and windsurfing will be discussed together here. When, however, it comes to discussing specific effects and solutions, it will at times be necessary to differentiate between them.

We can distinguish three kinds of facilities for sailboats according to size and construction:

a) Sets of buoys: moorings for boats on buoys. They are reached from shore by means of small ferryboats.

b) Docks: either single docks or groups consisting from a few to several docks. The docks are arranged in an open manner (that is, without a mole, with or without a breakwater).

c) Large marinas for sporting boats: large facilities for sailboats – and sometimes for motorboats – with surrounding walls of masonry or cement (concrete bulwarks, boulder piles, quay walls).

The last-named **marinas for sporting boats** can in turn have the following support facilities and amenities:

- Support facilities: halls for sheltering boats indoors, slips, cranes, repair wharf, sites for collecting trash, and used oil, sewage facilities, clubhouse with sanitary facilities, berths for mooring boats on land, automobile parking lots.
- Amenities: access roads (either new or widened pre-existent roads), recreational facilities (vacation village, tourist center), restaurants, yacht hotel (rarely found).

In the main marinas offer moorings to the following types of boats: sailboats, rowing boats, cabin cruisers, yachts, motor boats. Outside of marinas small portable boats, canoes, and surfboards are carried directly into the water.

When looking at the environmental stress that sailing and windsurfing can cause, we must distinguish at the outset between areas of activity connected to **land** (marinas, beaches, shoreline areas) and those on the open **water** (all waters used in sailing, anchorages). Sailing does not by necessity require marinas, as small, light craft can set out from accessible, flat shoreline areas.

Sites used by windsurfers often have simple facilities and amenities on land, such as

- places for setting down boards and rigging,
- stands for boards and gear with lockers,
- a lawn for sunbathing,
- toilets, trash containers,
- a parking lot with driveway.

2. Potential Conflicts

The following stress on the environment, arranged according to cause, can arise from:

- **Building a marina**

  Marina buildings on land, excavating and filling, paving and covering surfaces, bank or shoreline revetment, dredging channels for boats, and constructing the harbor basin, can affect the balance of nature and the quality of the landscape in the following ways (even when the above are present only in the construction period):
  - They can lead to the loss of valuable terrestrial, amphibian, and aquatic habitats (riparian woodland, wetlands, reed beds, water surface) or reduce their ecological quality.
  - They may alter or destroy the natural morphology of shorelines (topography).
  - Dredging and sailing in the shallow water zone (a very valuable limnetic ecosystem) can reduce its ecological quality: stirring up sediment into the water, sedimentation, damage to underwater vegetation such as potamogeton communities.
  - Barriers erected in the water can disrupt current patterns on their lee side causing areas of "dead water" (lack of currents with consequent insufficient oxygen flow produces areas of sludge), nutrient and sediment deposits, damage to vegetation, reduced water quality.
  - Facilities with visible machinery and vehicles can mar an otherwise natural, intact shoreline landscape.

- **Setting up a field of buoys can bring the following effects:**
  - Slips and winches for boats anchored with concrete (insofar as floating boat moorings have this equipment) can damage the natural morphology of shorelines as well as shoreline habitats.
  - The continuous movement of buoy chains in the water can damage underwater vegetation growing near the points where chains are anchored onto the lake or sea bottom. This is particularly a problem when buoy moorings are located in very valuable shallow water zones.
  - The exposed location of buoy fields can frighten or disturb birds that live on the surface of the water.
  - Buoy fields can mar the appearance of the landscape (insofar as the affected body of water is in a relatively untouched state).

- **Water sports activities (sailing and windsurfing)**
  Windsurfers can cause greater stress to animal and plant species (in particular, by disturbing waterfowl in reed areas near the shore and damaging vegetation with floating leaves) than sail-boaters, since it is easy to enter into shallow-water areas with a surfboard.

Areas of shallow water in particular, but other areas of water as well, can have great importance as habitat for waterfowl (for breeding, searching for food, and resting). Sailing or surfing in such waters can potentially:

- Stress, disturb, or chase away bird species which need protection and are susceptible to disturbance (by intruding into the animals' required safety distance).[20]
- Cause pollution to waters when sewage or oil are improperly disposed of, or when harmful substances (such as antifouling agents) are used to treat or clean boats.
- Tear or otherwise damage vegetation and stir up sediments: Sports activities in sensitive areas may push back valuable shoreline vegetation (such as reed beds and communities of plants with floating leaves) or break it into isolated patches, as well as harm water quality in shallow areas.

- **Secondary or subsequent effects of marinas**
  - Generating automobile traffic (from boaters and visitors), possibly entailing the construction of new access roads.
  - Construction of additional recreational facilities and amenities (tourist cafes, camper sites, and the like) shoreline zones of high ecological value.
  - Increase in water sports activities through expansion of existing marinas.

From an environmental point of view the construction of a sailboat marina can also bring positive effects, namely when the marina is part of an environmental protection plan for an area. Such a marina would obviate the need for buoy fields in the area and would act to concentrate sailing activity in unproblematic waters and shoreline zones.

---

20 One should note that the extent of this safety distance depends to a great degree on whether hunting waterfowl is permitted in the surrounding area, for hunting makes waterfowl wary.

3. Finding Solutions

The following measures and policies may serve as starting points in solving conflicts between sailing sports and the needs of nature and the environment for protection:

a) Demarcating zones for animals and plants that will be off-limits to sports activities

b) Siting marinas in environmentally compatible locations, operating marinas and other water sports sites in a manner that safeguards the environment

c) Preventing toxic substances from entering the water

d) Sailboaters and windsurfers showing consideration for animal and plant communities.

A) Off-limits areas

The specific ability of sailboating and windsurfing to disturb wildlife lies in the lack of a "set course" in these sports. The erratic and unpredictable movements of sailboaters and windsurfers increase the safety distance that birds need around them and hinder their ability to adapt. Waterfowl are designed by nature to flee quickly. Boaters and windsurfers can, however, act to modify the disposition of waterfowl to take quick flight. When over a sufficiently long period of time they sail or windsurf along the same routes and do not cross into off-limits areas, birds can gain confidence that their safety distances will not be violated. This permits waterfowl to decrease their safety distance on their own initiative, and even to swim towards boats. Thus the critical distances between water recreationists and waterfowl may be reduced after a moderate period of time. When, however, planners allot too little space for the birds' safety zone, shy species will leave and avoid the area while those more tolerant of disturbance will stay and perhaps even draw advantages from the situation. Yet, the goal of protection regulations cannot be to sacrifice threatened and rare species in favor of those that are already plentiful.

Sail-boaters and windsurfers disturb birds primarily in those areas where they breed, molt, and rest in the course of their migrations. The critical safety distances vary according to such factors as species characteristics, time of year, and pressure from hunting.

What do these off-limits areas look like, which planners should demarcate based on safety distances in order to guarantee waterfowl the necessary sphere of quiet that they need for survival? Which shorelines or water surfaces are especially suitable for these zones? We must consider the following areas having excellent ecological qualities:

- Reed beds more than 15 m wide and at least 300 m long, particularly when they lie next to aggradation zones, such as marshes and swamp forests
- River deltas more than 50 ha in size (regardless of whether rivers flow into natural and artificial lakes or the sea)
- Tidal flats (such as those found in the Wadden Sea National Park on the North Sea coast)
- Shallow bays and inlets with abundant underwater vegetation and a maximum depth of 5 m (the depth to which many species can dive)
- Small, uninhabited islands with trees used by birds for breeding
- Islands of various sizes having natural shorelines with reed beds and other plant communities growing in aggradation areas
- Island areas in dam reservoirs, particularly when these islands have interior lagoons.

Less valuable areas are shorelines and river banks altered by construction, steep shorelines (typical of dam reservoirs and mountain lakes), shorelines with reed beds less than 5 m wide, and shorelines with roads or railroad tracks.

These zones or habitats on lakeshores, in dam reservoirs, and along larger watercourses have, in turn, the following components of special importance:

- Reed beds and other tall reed communities used as areas for breeding or retreat (offering vegetative cover)
- Islands (with natural or near-natural vegetation) having the same functions
- Sandbanks and gravelly banks affording very specialized bird species (such as terns and little ringed plovers) places to rest, search for food, and breed
- Tidal areas and other areas of silt and mud subject to varying water levels that provide food for waterfowl.

Such sensitive zones as these are easily recognized and when necessary can be marked off with buoys. They form the core areas for demarcating quiet zones and their attendant safety distances.

The location and extent of off-limits areas (including buffer zones) cannot be determined at the local level alone. The significance of a given area within the larger whole must be considered and taken into account by planners at the federal, state, and regional levels. To help in this task conservation agencies have identified "Wetlands of International Importance" as part of the RAMSAR convention.

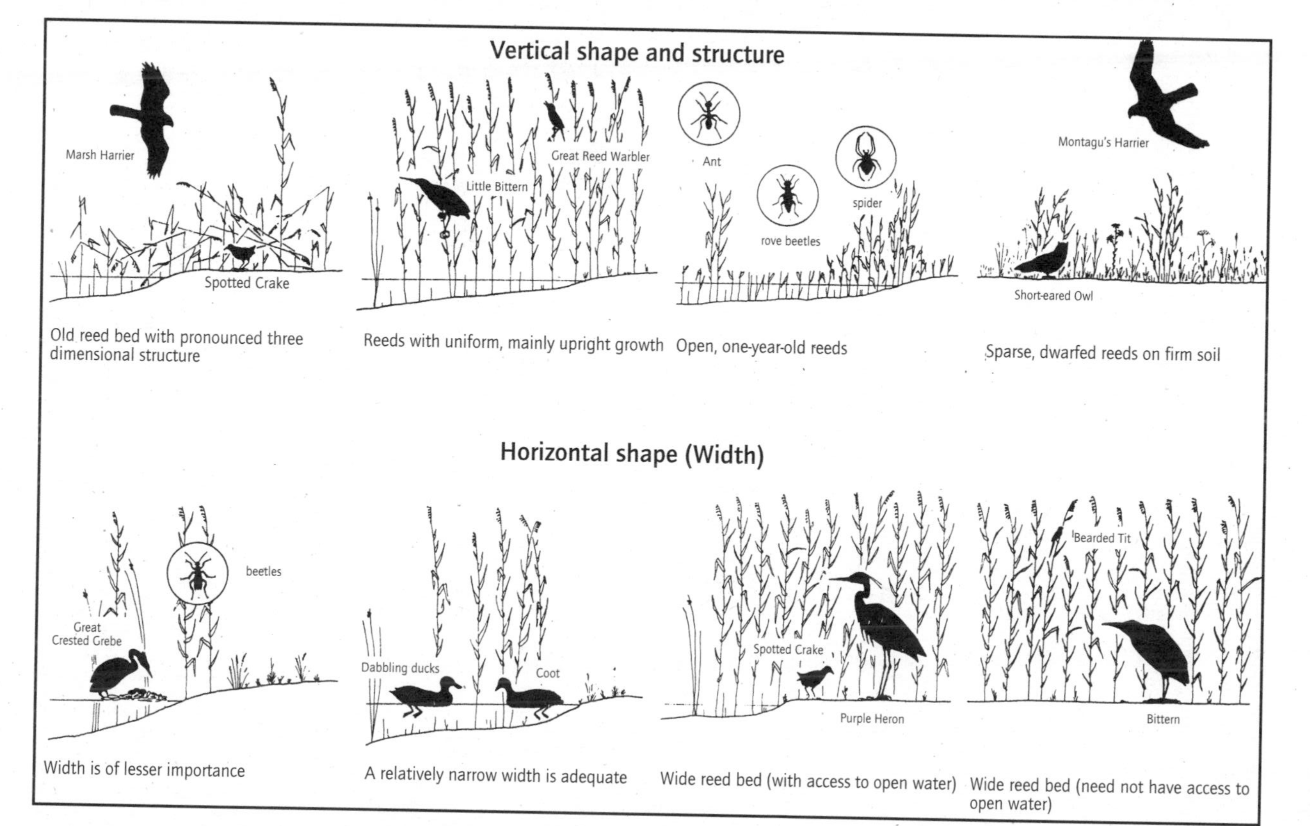

Reed bed structural features that determine colonisation, together with characteristic species

B) Siting and operating a marina

Marinas should only be built in shoreline areas already placed under a significant amount of stress, preferably industrial areas, cargo ports, and shipyards no longer in use ("land recycling").

Reusing former commercial areas spares open, unbuilt land from development. A marina in such a zone can make use of already existing infrastructure and services (supply and disposal systems, public transportation, parking spaces and shopping opportunities). Patrons could also find additional parking spaces in the vicinity after business hours and on weekends.

When possible, a marina should be located near existing recreational centers so that it can profit from their infrastructure, transportation-related and otherwise. When marinas are sited on already built-up and denaturalized shorelines, the additional stress that they place on the surrounding environment can be held within tolerable bounds.

C) Entry of toxic substances into water

Organic tin and copper compounds contained in antifouling agents can do significant damage to aquatic ecosystems.

It is possible that boats can do completely without antifouling treatment when they are taken out of the water several times in a season, or where the water harbors few organisms able to grow on the hulls of boats. The same holds true when aquatic growth can be relatively easily wiped, scraped, or scrubbed from boats. Besides the frequency of cleaning, the ability of organisms to adhere to boat surfaces and the type of water (fresh or salt) also plays a role.

The following measures are recommended:

- Do not apply coatings containing biocides to underwater parts of boats. A brush or a stream of pressurized water can remove the slimy growth that appears on boats kept in fresh water. Pressurized-water guns with more than 25-30 bars will, however, also remove paint. Do not use wood preservatives containing fungicides.
- When applying antifouling agents, place coverings on top of the surrounding area to protect soil, water, and plants from coming into contact with these toxic chemicals. Do not allow aerosol sprays to spread mist.
- When removing old paint, avoid contaminating soil and water. When sanding use sanding machines that vacuum up the scraped-off dust. Water from rinsing off abraded material or from wet-sanding must be collected.

- According to the German Federal Waste Avoidance and Management Act, sanding dust and antifouling residues are toxic wastes. Used brushes and paint rollers should be collected and disposed of separately from ordinary household garbage.
- When washing boats coated with antifouling paints containing biocides: do not allow the runoff water to flow back into the lake, sea, or river, nor into the sewage system. It must be disposed of as toxic waste or allowed to degrade until it has lost its toxicity. Reuse rinse water by setting up a "closed circuit system."

D) Behavior of sports enthusiasts

A successful environmental education helps people to translate their awareness of environmental issues and problems into concrete actions. Education about the environment complements other measures used to guide behavior, such as setting up signs, and disseminating information on proper behavior.

Behavior on the part of sports enthusiasts (and of windsurfers in particular) that respects the environment can help keep water sports activities out of ecologically valuable shallow water areas and zones of aquatic vegetation (reed beds, vegetation with leaves floating on surface, submersed plants). Such behavior also means maintaining sufficient distance (by staying out of animals' safety sphere and buffer zones) from the habitats of sensitive waterfowl, seals, and other species.

Warning signs, buoys, and floating barriers can make it clear which areas are off-limits to sports activities.

The following rules should be observed:

- Do not enter tall reed communities, reed belts and other sensitive shoreline habitats of birds, fish, and small animals from either land or water. The same applies to shallow waters (which may contain spawning grounds). Keep a sufficient distance from wading birds and waterfowl (determined by the size of the required safety distance, by buffer zones, and by the season of the year). On wide rivers, for example, this distance amounts to 30 to 50 meters.[21]
  When landing use only places specifically provided for that purpose or areas that are clearly not susceptible to disturbance.

---

21 Elsewhere a distance of 100 m for lakes is recommended.

- Do not set ashore or walk onto gravelly banks, sandbanks, or muddy areas where birds rest or roost, and do not approach seals' resting grounds in areas of tidal flats. Keep at least 300 to 500 m away from areas where seals or birds congregate; sail slowly past such places, and do not move out of waters marked for sailing. Observe and photograph animals only from a distance.
- In nature reserves and in relatively small lakes: Observe special rules governing sailing in nature reserves. In such areas sailing may be prohibited year-round or at least during certain seasons. Lakes under 10 ha in size should be closed altogether to sailing, unless such a lake has to be crossed in the course of a sailing tour on a watercourse.

MOTORBOATING AND WATERSKIING (R)

1. Significance for the Environment

Motorboating and waterskiing will be treated together here since the effects both activities have on the environment partially overlap (motorboats serve both as sports vehicles in their own right and pull water-skiers across the water).

The following categories of **motorboats** (or boats supplied with motors) exist:

- Motorboats with outboard motors
- Motorboats with inboard motors
- Sailboats with auxiliary motors (either outboard or inboard).

We can distinguish the boats' recreational use as follows:

- Motorboats used for pulling water-skiers and used as racing boats
- Motorboats used as sport or recreational vehicles without living quarters
- Motorboats with living quarters used for trips or cruising ("water tourism").

Waterskiing from motorboats takes place only to a relatively small degree, in areas designated for that purpose on large rivers, lakes and on coasts. The sport is mainly practiced in stationary water-ski facilities (without a boat). These facilities are normally located on lakes found in excavation pits. The lakes need not be larger than 4-5 ha. Pylons can be installed on the shore when the lake has a width up to 400 m. For safety reasons, the area taken up by skiing must lie at least 40 m from the shore or other parts of the water where activities are present. The water ski cables are driven by electric power.

A competition-level waterskiing facility has the following infrastructure and equipment (a facility in Paderborn serves here as an example):

- on the water: buoys, a ski jump (both anchored), two moorages for boats, a fueling station on the dock.
- on land: three towers where judges can view events, one boathouse (7x3 m) and cables with supporting pylons.

A new water sports vehicle has become popular in the last few years: the water **motorcycle**.[22] In contrast to motorboats this sports vehicle does not for the most part use propellers to move through the water, but rather a jet of water. Three categories of jet boats are allowed in racing: ski division (where the driver stands), sport division (where the driver sits) and runabout division (seating for two or more people). The high noise level produced when water motorcycles are driven at high speeds (or with engines buzzing and whining intermittently) has an irritating effect on people seeking recreation or residing in the vicinity. The danger they pose to swimmers and the high waves they produce are additional negative effects.

22 Also called a jet boat, aqua bike, jetbike, jet-ski, wetbike, waterscooter or waterbob.

2. Potential Conflicts

Stress on the environment from motorboats stems principally from the entry of toxic substances from motors into the water, from noise, and from high waves in shoreline areas. Other forms of stress, such as contamination from antifouling and cleaning agents, disturbing animals and causing damage to plants, also come about from sailboating and are thus discussed in chapter Q.

Toxic substances from boat motors that enter the water consist of hydrocarbons from unburned lubricants and fuel components. Any of their products that cannot be broken down in the environment or that are released by combustion also figure here. While the light hydrocarbons are relatively quickly broken down, the heavier elements sink down in lakes, sometimes all the way to the bottom. Their rate of decomposition depends on the temperature and oxygen content of the water. Even very small concentrations of petroleum-based hydrocarbons (in the ppb range: parts per billion) have a toxic effect on aquatic organisms. Yet even amounts below this concentration can lead to chronic poisoning or, by disrupting physiological processes, upset the entire ecological structure of the lake.

Because of their very high emissions, the usual two-cylinder motors found on boats (without electronic fuel injection and catalytic converters) are considerably more problematic for the environment than four-cylinder motors.

Noise: 75% of the outboard motors used in recreational boats have a performance of less than 10 horsepower (HP). The noise level of boat motors having a performance of under 10 kW (13 HP) lies between 62 and 70 dB(A) at a distance of 25 m. It lies between 70 and 77 dB(A) for motors over 10 kW. These measurements, however, hold only for professionally installed motors. Poorly mounted motors can produce twice the amount of noise [an increase up to ca. 10 dB(A)]. Differences in a boat's position on the water can cause fluctuations of up to 7 dB(A). Emissions levels can vary by up to 10 dB(A) within a single performance class of outboard motors, which indicates that considerable technological possibilities for decreasing noise exist (UBA 1988).

Noise level depends to a great degree on the speed of boats: the higher the number of engine revolutions, the higher the noise level. Limits on speed can thus considerably reduce noise emissions.

Damage brought about by **waves:** High waves coming from a motorboat's prow and stern and from water that has been agitated by waterskiing knock over weakened reeds or young plants and tear them loose. This occurs chiefly when motorboats or water-skiers venture too close to natural shorelines and when boats are travelling at high speed.

The stresses that motorboats place on animal habitats are like those brought on by sailboats and windsurfers (see chapter Q). In comparison to sailboats, motorboats can cause greater disturbance because of the higher speeds at which they can travel, and because of their ability to move about agilely and unpredictably (combined with noise). In areas sensitive to disturbance these characteristics of motorboats cause waterfowl to react with panic and so increase their safety distances.

A too small distance between a boat's propeller and the stream- or lakebed (less than 3 m) **will stir up sediment**. This has a negative effect on underwater habitats. The majority of Germany's federal waterways have a depth sufficient enough to preclude this problem.

This handbook will not consider the negative **social** effects of motorized water sport (for example, when water-skiers' activity drives away other people seeking to recreate in the same area).

3. Finding Solutions

From an environmental point of view motorboating is problematic if it harms aquatic habitats and/or significantly limits recreational activities oriented towards nature and the landscape. This is normally the case when:

- Noise from motorboats mars recreational areas requiring peace and quiet.
- People disturb sensitive aquatic habitats (particularly those used by waterfowl) by boating in or near them (thus intruding into animals' safety sphere).
- People boat in relatively shallow waters: this stirs up sediments, which in turn has serious negative effects on underwater vegetation and aquatic fauna.
- Boaters pass within 100 m of natural shorelines (having reeds and vegetation with floating leaves): destruction of vegetation from waves.
- People boat in waters having a high level of purity (the contamination of inland lakes with toxic substances is a particular problem).

Such areas must be entirely **off-limits** to motorboating. When doubt exists whether the conditions outlined above are present in a given area, the conservation authorities responsible for that area must be called in to decide.

We can distinguish two fields where measures can be introduced **to limit the amount of stress motorboating places on the environment:**

a) **Equipping and operating motorboats to reduce noise and exhaust emissions**

- Choose the right propeller (one appropriate for the boat and motor) to keep noise down as far as possible.
- Install technically advanced vibration absorbers in order to prevent motor vibrations from transferring to the boat. Modern outboard motors already have such absorbers positioned between the motor and the motor mount.
- Make sure inboard motors are encased in sound insulation (this can be added to already existing motors).
- Install mufflers on outboard and inboard motor exhaust systems.
- Mount an outboard motor so that it is vertical to the water surface (the angle of the motor relative to the water is adjustable).
- Check propellers as even light damage to them brings increased vibrations (and increased noise levels).

- Use liquid natural gas for motors (it reduces emissions and gets around the problem of leaking oil).
- Install a catch pan to collect oil leaking from the motor into the boat hull. Otherwise, the boat's water pumps will expel this oil into the lake or stream.
- Use only biodegradable lubricants on two-cylinder outboard motors.
- Equip two-cylinder motors with electronic fuel injection and catalytic converters.
- Use four-cylinder motors with catalytic converters.
- Operate motors properly and service them regularly. (Do not let motors idle, change motor oil and oil filters every 100 hours of operation, check water removal systems every 50 operating hours, and dispose of bilge water appropriately.)
- Do not paint the underwater part of the hull with antifouling agents that are harmful to aquatic organisms.

b) Motorboaters' behavior

- Before setting out on an excursion: familiarize yourself with the protection regulations in force for the lake or river you will travel on, and observe rules of proper behavior.
- Before letting the boat into the water: remove dirt, deposits of grease, oil, and other substances harmful to lakes and streams from the boat using biodegradable cleaning agents.
- Limit speed to 15 kph when near shorelines with natural vegetation (reeds and plants with floating leaves) to avoid damaging it and sending strong currents into sensitive areas.
- Dispose of trash, used motor oil, dirty bilge water, and the contents of chemical toilets in the places provided for their disposal. Before wastes may be properly disposed of, they must be stored in containers intended for that purpose.

FISHING (S)

1. Significance for the Environment

Fishing is a traditional recreational hobby that grew out of fishing as a source of livelihood. It enjoys popularity with an increasing segment of the population.[23] The effects of industrialization, in particular the pollution of waters and construction that has altered the banks and shores of watercourses and lakes, has, in comparison to earlier times, reduced opportunities for fishing.

Fishing differs from other activities treated in this handbook. The person who engages in this activity is bound or permitted (as the case may be) by law to protect fishing waters and use them in a sustainable manner. Thus the qualitative demands placed on fishing regulations have to be particularly high, especially as aquatic habitat is relatively scarce and reacts quite sensitively to ecological stress.

23 Fishing will be treated here as a recreational activity, not as a sport.

2. Conflicts

The following list of potential conflicts, which are not arranged in order of importance, will make clear the types of ecological stress that sport fishing entails. Many of these conflicts, particularly those that affect fish species, can be avoided through responsible measures of care and protection.

Stresses arising from **stocking lakes and streams with fish and from selective fishing:**

- Altering the existent species composition in bodies of water through releasing non-native species into them.
- Driving out autochthonous (native) species with allochthonous (foreign) ones.
- Introducing new fish diseases and parasites.
- Changing the genetic makeup of fish species by stocking streams with individuals originating in different watersheds.
- Bringing about a situation where non-usable small fish (hatchlings), amphibians (larvae and spawn), and water insects will be threatened by predators and competition for food.
- Lack of food and oxygen resulting from overstocking.
- Disrupting the food chain as a consequence of fishing for predatory species of fish, which are particularly attractive to fishermen.
- Using organic substances to lure fish when fishing in natural lakes and streams: the introduction of these substances into waters can lead to their eutrophication and deoxygenation.

Stress arising from **use of banks and shorelines** (docks, footpaths):

- Damage to vegetation on bank or shoreline
- Erosion of banks or shorelines.

Disturbing waterfowl while breeding, rearing chicks, molting (renewing plumage), feeding, resting, and sleeping, through:

- Blocking spaces suitable for breeding (when, starting in February and March, people fishing from shorelines or boats are present for long periods of time in such places).
- Driving away breeding birds from their nests: interrupting brooding in progress (when fishers remain for long periods of time in the vicinity of nests).

- Blocking spaces used for molting or causing disturbance in these spaces (molting birds unable to fly react particularly sensitively).
- Causing disturbance to flocks of waterfowl at sites they use for resting and feeding (while on their migrations in autumn or spring) and at wintering sites (consequently reducing the time waterfowl has to search for food).
- Endangering fish-eating species (such as cormorants and herons) by shooting them or using inappropriate means to keep them at a distance.

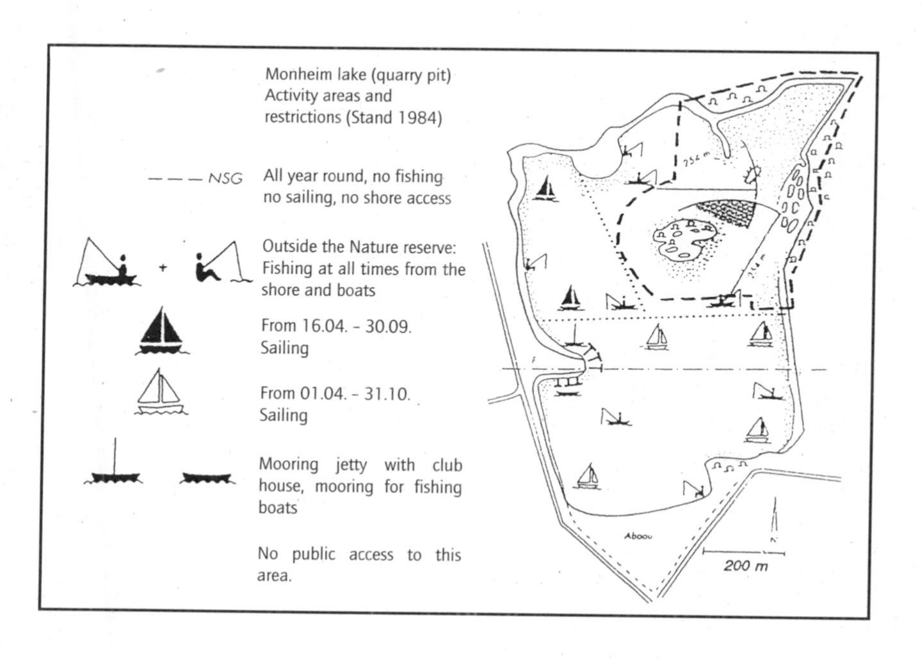

3. Finding Solutions

The following list provides an outline of measures that fishers can adopt as part of a policy of "protection and care." The measures will enable fishers to exercise their recreational activity in a way that safeguards the environment and even increases the ecological value of lakes and streams.

Protecting and caring for stocks of fish (without specific biotope or species protection)

- Monitor and control stocks of fish.
- Observe seasons closed to fishing (in order to protect certain species).
- Observe protective measures (catchable fish must have attained a minimum size).
- Establish standard mesh sizes for nets.
- Set aside areas where spawn must remain undisturbed (closed season).
- Place limits on catching fish (either on the number of fish that may be caught or on the fishing method used).
- Adjust the species composition and the amount of fish in a given stream or lake before stocking it with new fish.
- Aerate frozen streams and lakes (by making holes in the ice layer and keeping them open).

Combating pollution of lakes and streams

- Check water quality (biological and chemical analyses).
- Protest and lodge complaints against existing or planned sewage discharges.
- Remove trash and refuse.,

Taking a stand against other forms of interference in aquatic biotopes

- Oppose any bank stabilization measures (especially non-biological forms of construction).
- Oppose canalization and damming of streams in a close to natural condition.
- Oppose any clearance of shoreline vegetation and aquatic plants.
- Combat eutrophication that stems from farming near lakes (demand buffer zones).

Decreasing the effects of negative interference

Take part in:

- Restoring streams and lakes whose banks and shorelines have been altered through various forms of construction; protect bank and shoreline zones.
- Measures to protect stocks of fish in watercourses interrupted by weirs (channels for fish or fish ladders, eel ladders).
- Efforts to clear deposits of silt and mud (for example, around weirs).
- Efforts to keep sloughs connected to rivers where these have been dammed.
- Cut channels through man-made barriers in the streambed (so that fish may pass when the water level is low).
- Placing barriers before hydroelectric facilities to protect fish from harm (grills to keep fish out of dangerous areas).
- Removing tall, reinforced barriers in the streambed as well as any pipes or flumes through which streams have been made to flow.

Improving biotope diversity

- Develop and protect vegetation on shorelines and banks.
- Create sheltered places for fish to rest alongside slopes or by planting trees along the bank or shoreline.
- Create places for certain species of fish to spawn (spawning grounds).
- Construct underwater barriers on streambeds in a way that would allow resting places for fish to develop.
- Make shorelines or banks irregular when creating new bodies of water or re-configuring existing ones.

Protecting fish species

- Allow rare species of small fish time to regenerate, and safeguard their numbers (stocking streams and lakes with this goal in mind).
- Reintroduce once-native species to streams and lakes.
- Exercise restraint when stocking streams in order to protect rare small fish (hatchlings).
- Protect rare species of fish and work to increase their numbers regardless of their value as game fish.
- Place certain species of fish under year-round protection.
- Designate certain areas as fish protection zones (make them permanently off-limits to fishing).

Species protection in general

- Do not fish in nature reserves.
- Do not stock ecologically sensitive lakes and streams with fish.
- Exercise restraint when stocking streams in order to protect amphibians (spawn and larvae).
- Protect kingfishers and other threatened fish-eating species.
- Make certain areas off-limits (prohibiting fishers from entering them).
- Maintain seasonal limits on use (for example, during times when birds are breeding and molting).

Alpine Skiing (T)

1. Significance for the Environment

Skiing as a professional sport is classified as either "Alpine" or "Nordic." Alpine skiing comprises the disciplines downhill skiing, slalom, giant slalom (Alpine combined) and super giant slalom. Nordic skiing comprises cross-country skiing, ski jumping, Nordic combined, and biathlon. Other types of skiing are long-distance ski jumping, freestyle, and – increasingly in recent years – snowboarding with its associated disciplines also have a place here.

From an environmental view, skiing is most problematic when regarded as a mass phenomenon. For this reason we shall focus almost exclusively here on skiing as a recreational sport. Most attention will be given to skiing in high mountain ranges such as the **Alps**. Downhill skiing ("Alpine skiing") is of course also practiced in lower-elevation mountain ranges such as the Central German Uplands, though this activity plays a minor role here due to relatively unfavorable snow conditions. Yet even here a few resorts enjoy a well-developed winter skiing season.

The analysis and conclusions offered in this chapter also apply to a large extent to ski areas in mountain ranges of low elevation. Skiing, however, affects mountain ranges of lower elevation differently than it does the Alps. Specifically, these lower mountains have different animal and plant habitats than the Alps, and the sites where their ski slopes are located are less subject to erosion or disruption of their water regime. Lower-elevation mountain ranges will figure more largely in the discussion of cross-country skiing in chapter U.

The German Skiing Association (DSV) and CIPRA, the International Commission for Protection of the Alps, estimate that the entire Alpine massif has the following numbers of ski lifts and ski slopes (exact figures are not known): 11,500 lifts, and between 15,000 and 18,000 ski slopes having a total length between 18,000 and 25,000 km. Approximately 20 million skiers visit the Alps each winter season.

Methods for transporting skiers on slopes may be classified into two main groups:

- Aerial cableways or chair lifts: Passengers ride through the air either in enclosed cars or gondolas, or while sitting on open, individual seats.
- Ground-level ski lifts: A cable pulls passengers up the slope as they stand or lean against a metal bar or hold on to a handle.

Ski slopes are divided into marked slopes and routes and unmarked, natural slopes. A **ski run** is intended for downhill skiing and is readily accessible to skiers. Its width depends on the capacity of the adjoining cableway or lift to transport passengers. Such a slope has been prepared and marked for skiing, secured from various alpine hazards, and is regularly maintained. Ski runs are laid out and, when designers deem it necessary, constructed with earth-moving machinery. **A ski route** is also intended for skiing, is accessible to all skiers, and is marked and secured from dangerous hazards. A route, however, has not been laid out or prepared specifically for skiing, nor is it maintained. **Natural** or "wild" **slopes** are ski areas accessible to the public, but are neither marked nor maintained in any way. On these slopes skiers thus pass through areas of deep snow located away from marked runs and routes. Those who ski here, however (as opposed to cross-country skiers), make use of cableways and lifts.

Possible Infrastructure in a Ski Resort

Possible Infrastructure in a Ski Resort	
Traffic Access roads Service roads Parking lots **Lifts** Stations on summits and in valleys Cable cars, chair lifts, ski lifts, cars for carrying equipment Entry and exit points for chair and ski lifts Buildings for lift machinery and personnel **Food and Drink** Restaurants Mountaintop cafes/restaurants and mountain cabins with outdoor terraces Small refreshment stands located in dairy sheds in high Alpine meadows "Snow pubs" **Utility Supply and Waste Disposal** Water pipes Electric and telephone lines Sewer lines Septic pits or tanks **Ski Slopes** Downhill slopes	Ski tracks linking slopes Safety fencing Snow fencing Areas for dumping excess snow Drainage ditches Walls for supporting steep slopes Protective barriers in steep creek beds Floodlighting Plastic matting **Snowmaking Machinery** Snow cannons Pumps for drawing water Pipes Water storage tanks Cooling towers Buildings housing equipment for obtaining water **Avalanche Protection** Open corridors reserved for blasting avalanches Avalanche barricades **Other Infrastructure** Toboggan runs Summer tobogganing tracks Cross-country skiing tracks

Source: Leicht et al., 1993

In recent years several new **trends in winter sports** have drawn the attention of the media, even though the number of people engaging in these activities (such as carving, "big foot," freestyle, and trick skiing) is usually very small. **Snowboarding**, however, has moved from a sideline phenomenon to a strong trend over the last few years. This new sport, moreover, has importance from an environmental point of view, since snowboarders do not limit their activity to slopes specially prepared for snowboarding (half pipes and fun parks), but to a great extent search out areas of deep snow away from ski slopes ("powdering"). When these natural areas provide habitat for species needing protection, visitor management measures become necessary from an ecological point of view.

2. Potential Conflicts

The following section lists negative impacts on the environment that arise from construction and operations taking place in ski areas. The forms of stress are arranged according to the activity that causes them.

- **Construction and lay out of technical infrastructure** (buildings, cableways and lifts with supporting pylons and their foundations, trenches for laying cables, facilities for utilities and waste disposal, roadways, parking lots). These forms of interference in the environment, though mostly limited to small areas, do not differ much from other construction activities (such as the paving and covering of surfaces, introducing radical changes in existing ecological relationships, in particular in animal and plant communities) in their significance for the environment. They do, however, have unique characteristics in the following areas:
  - Waste disposal: Treating sewage and disposing of garbage without causing harm to the environment is very cost-intensive.
  - Marring of the landscape: Pylons and other forms of technical construction in the mountains are often visible over a great distance.
  - Erosion problems: Because of the extreme elevation changes found in mountain areas, even minimal disturbance to the ground can lead to erosion.

- **Construction of cableways and lifts**
  - Felling trees, cutting a corridor through forests
  - Grading the land (when building a ground-level ski lift).

- **Construction of new ski runs, expansion of existing ones**
  - Clear-cutting of trees in mountain forests (to make corridors for ski runs). Consequent loss of functions performed by the forest: preventing erosion, maintaining water balance, providing habitat for animals and plants, preventing avalanches. Damage from heat and wind can occur along exposed forest edges.
  - Removing individual trees: When roots are dug out, the soil lies exposed and is open to erosion.
  - Clearing fields of dwarf pine: Protective functions as well as habitat are lost.
  - Grading land (planing, dynamiting rock, altering incline of slopes by cutting and filling, flattening steep inclines, filling in hollows): Exposing the soil and altering its structure, destroying or altering vegetation cover with consequences for erosion, surface drainage, species diversity.

- Destruction of valuable biotopes, such as wetlands; small streams; alpine meadows on uneven terrain; and habitats of rare species.
- Disturbing game animals and driving them away from their customary grazing grounds located near ski slopes (resulting in damage to forests when animals then feed on buds and bark there).
- Marring the landscape (from graded slopes showing marks of erosion and having untypical vegetation).

- **Maintenance of ski runs**
  - Noise from large machinery and snow cannons (see section 4 for snowmaking machinery)
  - Exposing and damaging the humus layer (from using ski-run machinery when ground is not sufficiently covered by snow)
  - Damaging turf, dwarf shrubs and stands of dwarf pine (from using machinery when ground is not sufficiently covered by snow)
  - Reducing the growing season (by compacting snow, by causing ice to build up, and from artificial snow), and causing the eutrophication of mountain lakes (from fertilizer runoff)
  - Altering plant communities and/or losing species diversity due to eutrophication of nutrient-poor soils (resulting, for example, from artificial snow). On snow making see the excursus, section 4.

- **Skiing**
  - Damage to vegetation when ground is not sufficiently covered by snow (from edges of skis).
  - Skiers using natural areas away from prepared ski runs can disturb sensitive species (in particular, grouse). The increasingly simplified design of ski slopes (through grading) leads more people to ski in natural areas.
  - Stress on vegetation located away from ski runs caused by skiers using natural areas.

In order to make a year-round ski season out of the relatively short winter season, ski resorts have been placed in **glacier regions** (on approximately 30 Alpine glaciers). Glacier skiing poses dangers to future supplies of drinking water, for the Alpine glaciers constitute important reservoirs of drinking water for Central Europe. A great variety of substances (such as wax and oils of all kinds) are released by the preparation and operation of glacier ski runs as well as in restaurants and cable car stations. Sewage and wastes that are practically unavoidable or can only be disposed of with great difficulty are also produced. When these environmentally harmful substances are not properly disposed of they become "deep frozen" and are then released decades later. Thus future generations must bear the consequences of today's glacier pollution.

Skiing and snowboarding in areas away from prepared slopes also leads to ecological problems when grouse are driven away.

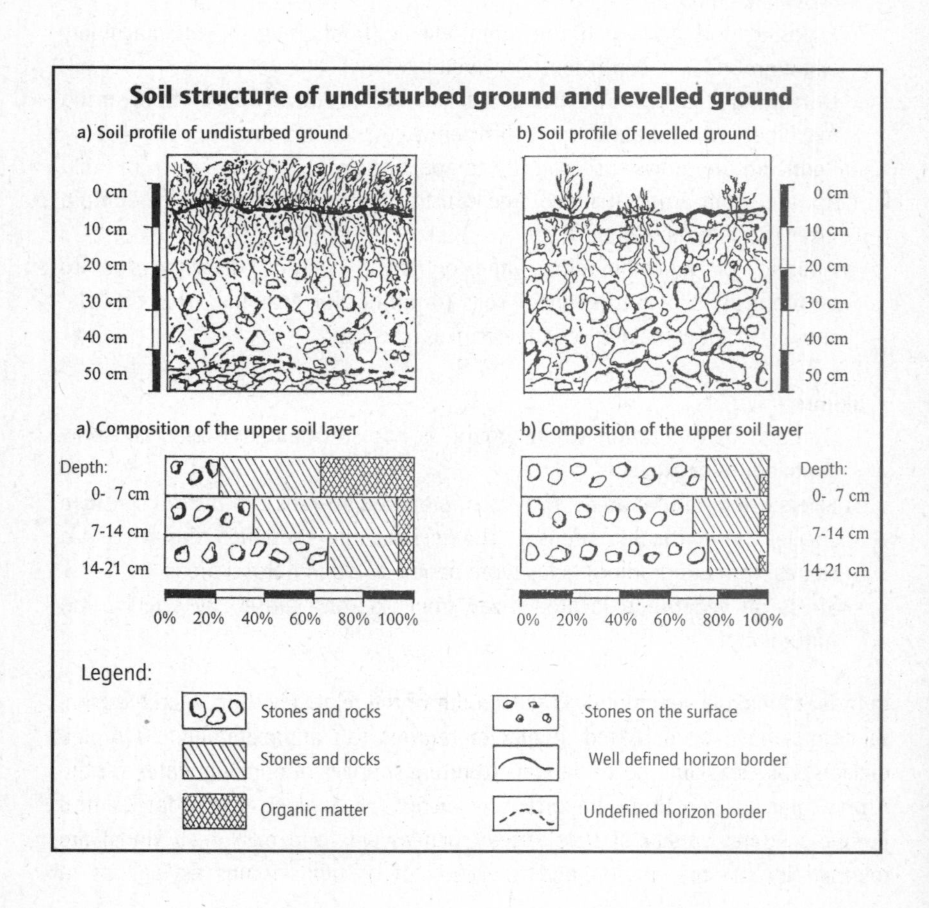

We must also consider stress on the environment that results from **travel by automobile.** In comparison to other recreational sports, skiing is particularly affected by the automobile and its associated problems

- because, as a recreational destination, the Alps are marked by a particular ecological sensitivity (key terms and concepts here: dying forests (Waldsterben), danger of environmental catastrophe from breakdown of mountain forests, auto emissions as the principal cause of both, see chapter A),
- because skiing as a mass phenomenon in the Alps constitutes the greatest "engine" of winter tourism, particularly at weekends.

According to a study by CIPRA, tourists drive between 15 to 25 million kilometers a year in the Alps. (This figure includes distances traveled to and from an Alpine destination both by people making day trips and by those taking a longer vacation, as well as additional trips made by vacationers during their stay.) This amounts to approximately 20% of the total traffic moving on Alpine roads.[24] Which percentage of this is due to skiers was not surveyed.

Among those forms of environmental interference outlined above, the **grading of slopes** must be regarded as the most serious ecological problem by far (next to the cutting of forests, which hardly occurs today).

We can distinguish the following complexes of problems resulting from grading slopes to make ski runs:

- Harm to the protective functions performed by soil and vegetation (decreased water retention capacity, triggering various forms of erosion)
- Harm to the habitat function of the original area (whether this was pristine, near-natural, or extensively used) with respect to the number and variety of ecological niches
- Marring the visual beauty of the landscape.

---

24 Residents generate 70% of the total 100 billion kilometers driven each year in the Alps, while transit traffic accounts for 10% of it. (Gruber 1994)

3. Finding Solutions

Constructing ski runs

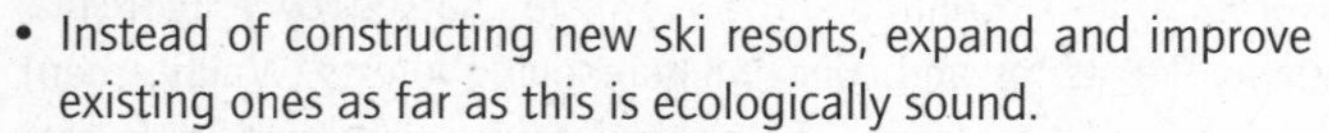

- Instead of constructing new ski resorts, expand and improve existing ones as far as this is ecologically sound.
- Adjust the passenger capacity of aerial cableways and lifts to match the carrying capacity of ski slopes (not the other way around).
- As far as possible, let ski slopes retain their natural condition (instead of creating wide, even "skiing freeways").
- Do not plane large surface areas (slope grading), instead make small, individual adjustments to the slope, and only then in exceptional cases. Do not plane above the tree line.
- Valuable biotopes may not be incorporated into ski slopes or lie next to them. Ski slopes that intrude into valuable biotopes must be reconfigured or taken out of operation.

Maintenance, restoration, and renaturalization of ski slopes

- Maintain (or prepare) slopes when sufficient snow cover lies on the ground in order to minimize damage to soil and vegetation. Skiing down slopes should only be permitted when snow entirely covers the ground.
- Difficult terrain and areas where deep snow collects should be left intact when constructing (or preparing) ski slopes. This will make the slopes attractive to skiers searching for natural, "wild" areas to ski in.
- Use snow cannons (artificial snow): see excursus.
- Measures to restore vegetation to mechanically altered slopes and to return them to their natural state must be planned and carried out with great care making use of the latest scientific knowledge – even if this means a great deal of effort. Measures should promote diversity of vegetation and aim to restore a diversity of niches for small animals (uneven terrain, "micro-relief").
- Restore damaged slopes using the latest techniques for the greening of ski slopes. The aim here is to return the area to a natural condition.
- The following points are essential for ski resorts located in areas with glaciers, which are particularly sensitive: use biodegradable oils in machinery, use waxes free of harmful substances, limit applications of salt, and take pains to dispose of solid waste and sewage properly.

Skiers' behavior

The German Ski Association (DSV) addresses "the skier in nature" with the following rules on behavior: "Skiers experience nature at first hand. It is home for animals and plants that thrive on sensitive ground. Nature also protects mankind itself. Everyone is called upon to take care of the landscape so that people may be able to engage in winter sports in a natural environment in the future. The DSV thus asks skiers to please observe the following rules:

1. Stay on marked ski runs, tracks for cross-country skiing, or marked routes.
2. Do not leave prepared tracks in the forest and do not engage in deep snow skiing in the forest. You would disturb animals and damage the sensitive growth of young trees with your skis.
3. Ski only in areas completely covered with snow.
4. Do not ski in natural areas away from ski runs where such skiing is not expressly allowed.
5. Observe posted signs and stay clear of protected areas.
6. When skiing, leave your dog at home.
7. Avoid making noise.
8. Do not ski during twilight or at night.
9. Take all trash home with you.
10. When possible, use public transportation to travel to a ski site."

Another important recommendation is to practice other sports or activities in an environmentally sound way as an alternative to downhill skiing (ice skating, curling, sledding, hiking, cross-country skiing, and tour skiing).

Besides not driving to the ski site, the individual skier can make an important contribution toward limiting stress on the environment by not skiing on slopes with insufficient snow cover. Yet the responsibility here lies primarily with ski lift operators, local communities, and authorities charged with overseeing ski areas. They must see that cableways and lifts are shut down when insufficient snow lies on the ground. Only then can we effectively protect the vegetation cover on ski slopes from overuse.

4. Excursus: Snowmaking machines

We can distinguish the following purposes in snowmaking:

- Covering a large area with snow
- Applying snow to fill out small problem zones
- Creating a snow depot (a "supply hill" made of artificial snow which can then be applied to slopes by means of bulldozers).

To make snow, water, air, and energy are needed. In Germany and Austria chemical or bacterial additives are forbidden.[25] The man-made production of snow rests upon the following physical principles: fine water droplets are created by means of pressurized air and are then hurled out of a snow cannon. Some of these water droplets vaporize before they hit the ground, provided that the surrounding air is not saturated (humidity well below 100%). This vaporization process causes (according to physical laws) the water to lose heat. Thus the water droplets hurled out of the cannon freeze and turn into snow crystals before they touch the ground.

Artificial snow can be produced under certain atmospheric conditions. The following **conditions** must be met before snow can be produced:

- Air temperature colder than -3° C
- Humidity below 80%
- Water temperature colder than +2° C.

Normally, two processes are used to make snow: high-pressure machinery (pressurized air cannons), and low-pressure or propeller machines. Low-pressure machines are always mobile, while high-pressure machinery is normally anchored in one place on the edge of a ski slope.

Artificial snow differs from natural snow in the following ways:

- Artificial snow has a different crystal structure than natural snow not prepared by man. Artificial snow is more compact than natural snow. Air passes more easily through artificial snow and it holds in less warmth. Yet a snow cover

---

25 These ecologically problematic additives are mixed into the water so that snow can be produced at higher temperatures. Among these additives are dead bacteria (Snomax™) that cause water to freeze more quickly (forming ice crystals).

composed of artificial snow produced under favorable atmospheric conditions is comparable to a snow cover made by spreading natural snow with bulldozers onto a slope and compacting it.

- Because of its different crystal structure, artificial snow melts more slowly than natural snow. Depending on local conditions it can take up to 25 days longer for artificial snow to melt (Mosimann 1991).
- The surface water used in the production of artificial snow has a much higher mineral content than rainwater and also has a higher pH value (Kammer 1990). This water has thus a fertilizing effect that may be desirable in areas used as pasture. Yet it causes serious stress to plant communities that require a soil poor in nutrients.

Water usage varies according to the desired height of snow and the conditions prevailing at the time the snow is made. Usage lies between 80 and more than 200 liters per square meter in a winter season. To produce snow in sufficient quantities to completely cover an area and keep it covered, one must plan on using 200 liters for one square meter of snow (for a depth of 30 cm of compacted snow).[26] A medium-sized slope of 10 hectares thus requires approximately 20,000 m3 of water (Schatz 1990).

Because this water is not consumed, but reenters the water cycle when the snow melts, ecological stresses only occur

- when the withdrawal of water from a lake or stream causes damage to that body of water during times of naturally low water levels in winter.
- when the higher volume of meltwater (the sum total of melted artificial snow and subsequently fallen natural snow) contributes to erosion or flooding.

The availability and quality of water are the most critical aspects involved in snowmaking. Removing water from streams and lakes for snowmaking can place stress on the plant and animal communities living in them. Either too little water may be left causing areas to dry up or ice to form on the stream- or lakebed, or the continuity of a stream may be interrupted by damming in order to draw away water. The body of water from which the water is to be removed must remain

---

26 In extreme cases, when conditions during a particular winter are unfavorable, up to 600 liters per square meter may be needed (Mosimann 1991).

intact as habitat (preventing ecological problems). The lake or stream in question should be checked for the amount of sewage present, for it must maintain its ability to absorb wastes that flow into it. Moreover, this same lake or stream must have water of sufficient quality (with respect to nutrient levels and germs that endanger health) for use in snowmaking. Even when the water is free of germs and contaminants it has a different composition than water in the form of precipitation.

Sometimes ponds are constructed in order to store water needed for snowmaking. The aesthetic effect of such a newly constructed body of water and its viability as habitat are important here. Streams may not be dammed for this purpose.

The **energy consumption** of snowmaking machines varies from 2,000 to 27,000 kilowatt hours per hectare of land a year (average 13,000 kWh). Large machines consume about as much energy a year as a 50-bed hotel (Leicht 1993).

A word on **noise**: Snowmaking machines produce noise, which can be particularly annoying when machines run at night. Noise emissions from the loudest snow makers (high-pressure systems) lie between 91 and 99 dB(A), in the same range as jackhammers. Low-pressure systems are not as loud [58-80 dB(A)]. The machines must be placed at least 800 to 1,000 m away from a residential area in order to stay below the limits on noise set by law. Noise emissions can be reduced by at least 10 dB(A) in each type of machinery when the snow cannons are acoustically fine-tuned.

Kammer and Hegg (1990) studied ski slopes in Savognin, Switzerland (altitude 1,200 to 1,800 m) that for eight consecutive years had lain under artificial snow. They found that these slopes (compared to neighboring areas that had been covered only with natural snow) experienced the following ecological changes:

- The spectrum of vegetation shifts: Species typical of dry and nutrient-poor sites are pushed out by common species able to grow in a variety of habitats.
- The number of species per measure of land decreases sharply by a fourth to a third. This holds true for both nutrient-poor and nutrient-rich meadows as well as for subalpine grass areas.
- The percentage of grasses in the ground cover decreases substantially by between 17 and 33%. The soil is thus less knit together by roots, which makes slopes more prone to erosion.

From an ecological point of view, the most serious and direct environmental stress from artificial snow affects valuable oligotrophic sites (that is, areas characterized by nutrient-poor soil and dry conditions). Artificial snow destroys these sites over the medium term. Kammer and Hegg (1990) showed quite impressively how vegetation was trivialized and the number of species reduced. The authors reached the following conclusion from this: "From the viewpoint of ecology and nature conservation, snow making should be rejected categorically and without exception on nutrient-poor meadows and pastures, and on raised bogs, mires, and fens (and possibly on heaths of dwarf shrubs) (ibid.)."

The **indirect consequences** of snow making can be particularly grave when ski resort operators use snowmaking machines as a means to achieve the following:

- Extension of the ski season beyond its natural climatic limits.
- Overcoming a chronic lack of snow in ski areas having generally insufficient snow levels.
- Constructing new ski lifts and slopes, which in turn can only be profitably operated through the use of snowmaking machines.
- Altering the terrain of existing ski slopes that can only accommodate snowmaking machines when graded.

From an environmental point of view, snowmaking machinery is tolerable when it is used to ensure that a ski slope remains usable for skiers during the **normal winter season.**[27] Snowmaking machines thus may not be used to bring the beginning of the ski season forward or to extend it beyond its natural limits.

Snowmaking machines may not be used in the **ecologically sensitive parts** of areas where their construction and operation would entail considerable negative effects on nature and the landscape that could not be counterbalanced in some way. This is particularly true when resort operators plan to use snowmaking machines on key sections of the following types of ski slopes (Leicht 1993):

- Slopes above the tree line
- Slopes that to a great extent harbor near-natural plant communities (for example, nutrient-poor grasslands, bogs)

---

27 Climate change has also to be considered here. When average temperatures rise, we cannot use the colder temperatures that previously marked a winter season to define a "normal winter season."

- Slopes with insufficient vegetation and constant erosion problems (for example, where vegetation covers less than 70% of the surface and where soil is not knit tightly together by plant roots)
- Geologically unstable slopes that are constantly threatened by erosion
- Slopes where, in order to use snowmaking machines, it would first be necessary to fell trees and grade the surface
- Areas that are also key habitat for threatened, sensitive animal species active in winter (such as grouse).

In order to interact responsibly with nature and the landscape, an environmental impact assessment must be carried out before deciding "whether" and "how" such machinery may be employed.

Cross-country and Tour Skiing (U)

1. Significance for the Environment

Cross-country and tour skiing have a markedly different effect on the environment than alpine skiing on prepared and natural slopes. For these two sport activities do not require ski lifts,[28] much less a ski slope which in many cases had to be reshaped and cleared of trees before it could be used for winter sports.

Cross-country skiers use tracks made in the snow in mountain valleys, in lower-elevation mountain ranges, and on flat land, in equal measure. This "soft" and quiet sport, which scarcely needs any infrastructure, can nevertheless come into conflict with nature conservation when it is practiced in the wrong place. Special machinery makes tracks for cross-country skiing in the snow, which are marked, maintained and safeguarded from dangers (such as alpine hazards). If the skiers use a freestyle or "skating" technique (not compatible with the normal style of skiing on the same track), the track does not need to consist of two ruts in the snow into which the skis fit, but may simply be prepared with a roller.

Tour skiing is characterized by not being tied to pre-made tracks. Ski routes can, however, be marked. All parts of mountainous landscapes are potentially suited for tour skiing, as long as no natural barriers (such as very rough relief, areas subject to avalanches) or technical barriers (roads, houses) stand in the way.

28 Tour skiers, however, often use lifts to reach a more advantageous starting point for a tour.

2. Potential Conflicts

The potential environmental conflicts posed by these two "soft" forms of skiing are relatively minor compared to those from alpine skiing when we exclude secondary effects (such as use of motor vehicles).[29] Potential conflicts with nature conservation are limited to those few areas in which skiers disturb the habitat of plants and animals meriting protection. Other potential conflicts (such as damage to young trees in forest plantations or causing disturbance to hoofed game animals) are predominantly of an economic nature or negatively affect hunting interests.

The potential for disturbing rare species (particularly grouse) arises from activities that upset the affected animals or drive them away. When the animals cannot move to a suitable refuge the existence of entire populations can be threatened.

29 Questions involving consumption of resources (such as in selection of materials, use of energy) that also play a role here are discussed in chapter A.

3. Sensitive Habitats of Grouse Species

The main ecological problem posed by cross-country and tour skiing lies in the possible disturbance or driving away of grouse.
The following discussion will focus on the requirements and in particular on the sensitivity of black grouse, capercaillie, snow, and hazel grouse, all of which are seriously endangered (they exist only in small, isolated communities) and thus greatly in need of protection.

The **black grouse**[30] requires the following kinds of biotope in open landscapes free of disturbances:

- Areas with stable vegetation formed by the prevailing climatic conditions such as fens, bogs or areas near the tree line (natural biotopes)
- Heath-bog landscapes, extensively used meadows and pastureland, and low-elevation mountain ranges (biotopes marked by human uses).

In the Alps the black grouse lives at elevations above the dense forest belt, primarily in high mountain pastures located in the dwarf forest zone (composed of, for example, dwarf pines and green alders), in forests near the tree line, and in communities of dwarf shrubs. This grouse seeks out vegetation of differing heights (from pastures to high-growing vegetation in transitional areas to the edges of forests) for the different phases of its yearly or life cycle (breeding, rearing young, mating, searching for food, resting, and molting).

A black grouse population that is able to survive, consisting of approximately 50 individuals, requires an area of at least 1,500 hectares. This area must offer a suitable biotope, be largely free of disturbance, and be contiguous. The black grouse as a species is everywhere endangered. Relatively stable populations still live in the Alps. The principal cause for the great decrease in populations of black grouse are the changes in land use and landscape structure that have taken place in the last decades. The shrunken numbers of black grouse, found now in remnants of their former range, are also endangered by outside disturbance.

---

30 In German, black grouse are sometimes called "birch game." Capercaillies are also referred to as "game." The word "game" here shows that these animals were once hunted and so describes the hunter's point of view.

The most serious form of disturbance arises from recreational uses, that is, when a person seeking recreation (such as a skier) enters the grouse's habitat. Disturbance agitates birds (that is, it causes them to take cover or interrupts their normal routine, but does not cause them to leave their habitat), or leads them to abandon their traditional habitat and to seek out another, less suitable biotope. A human passing within just 200 to 300 meters of a black grouse will cause it to fly away.

The **snow grouse**, also a threatened species, lives predominantly in higher elevations of the Alps (between 1,800 and 2,400 m). It is also found in the foothills of the Alps (at sites between 1,400 and 1,600 m). Its "partners in conflict" are not cross-country skiers, but tour skiers (and skiers using natural, "wild" slopes, see the chapter "Alpine Skiing").

The capercaillie may be described as a characteristic species of mountain forests rich in coniferous trees. It exhibits a marked attachment to traditional mating, breeding, and wintering grounds. This species requires open, airy forests with rich and well-developed ground vegetation (with dwarf shrubs), where the birds seek food and cover.

The capercaillie is also seriously endangered. Small remnant populations occur in various low-elevation mountain ranges, somewhat larger ones in the high Alps and in the Alpine foothills. Capercaillies continue to decrease in number.

Low-intensity use of the forest for recreation is compatible with the habitat requirements of capercaillies, provided that visitors stay on marked paths or ski tracks and that any capercaillie biotope near these zones offers thick vegetative cover. Where, however, ground cover is sparse, a single human can cause disturbance to a capercaillie biotope from a great distance.

The **hazel grouse** lives in forests having a complex structure and admitting enough light for a rich blanket of perennials and shrubs (offering sufficient cover and food) to grow. Hazel grouse are sensitive to outside disturbance, but react with particular sensitivity to recreationists during breeding season and when rearing their chicks. A single pair of hazel grouse requires at least 30 ha of habitat (DSV 1995).

While grouse species are adapted to harsh winter conditions, they can be endangered when skiers' activities seriously limit their choice of winter habitat or the time they have available for their activities. The birds' reserves of energy can be overtaxed (that is, become fatally deficient) when they have to flee from danger or have only limited opportunities to feed.

4. Finding Solutions

a) Tour Skiing

In areas providing habitat for grouse tour skiers will have to be appropriately channeled through the area and existing routes may have to be altered in order to protect threatened grouse. Mandatory and enforced limits on skiing will be necessary where local grouse populations have wintering grounds that are essential to their survival. The German Alpinist Association, working together with representatives of sports and conservation organizations, carried out an information campaign entitled "Tour Skiing in an Environmentally Friendly Way" that introduced and explained necessary visitor management measures to tour skiers. The following principles formed the basis of this initiative:

- Ski touring should continue to be allowed in all larger regions.
- Touring routes that are particularly popular should remain intact (as far as is ecologically sound).
- Measures must be taken to ensure that grouse survive even extreme winters.
- Contiguous grouse populations may not be cut off from each other.

Maps have already been issued for some areas that show environmentally sound routes. Appeals were made to publishers of hiking maps to incorporate changes to touring routes into their maps.

b) Cross-country Skiing

In examining disturbances that may be caused by cross-country skiing, we have to distinguish between stresses

- that arise from a cross-country ski track that runs through a capercaillie biotope, and those
- that arise from cross-country skiers who leave prepared tracks and ski through natural terrain.

The single greatest demand on cross-country skiing is that tracks be laid out in such a way that natural areas are protected. Ski tracks may not cut through the habitat of grouse species. The same holds true for other ecologically valuable and sensitive areas. In areas under strict protection, planners will have to carefully choose routes for ski tracks or forgo them altogether in order to avoid placing stress on sensitive zones.

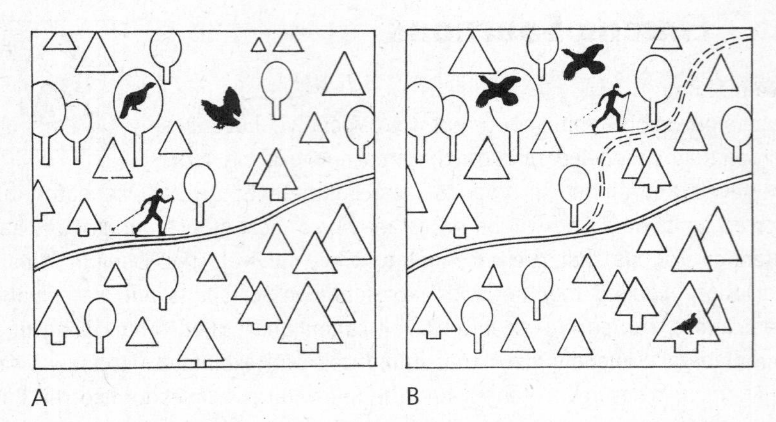

Grouse remain in their refuge areas undisturbed by skiing activities, as long as the ski routes lead past the sensitive areas and provided that cross-country skiers do not leave the route (A). Only skiers who leave the marked routes disturb the birds, forcing them to take flight (B).

In laying out ski tracks planners should also pay attention to the following:

- Linking tracks to the public transportation network: Tracks and systems of tracks should be physically connected to public transportation stops (train and bus), or should be easily reachable with public transportation via shuttle services.
- Infrastructure for cross-country skiers (restaurants, sanitary facilities, lockers, parking spaces) should be integrated into the existing infrastructure of villages and towns in the area.

Cross-country skiing tracks – like bicycle paths – constitute an important element in "soft tourism" and so deserve support from environmentalists and those engaged in nature conservation. For people can fulfill their needs for sport and exercise through these activities and – provided that they observe the measures outlined here for avoiding stress on the environment – protect the landscape at the same time.

9th Council of Europe Conference of Ministers responsible for Sport - "A clean and healthy sport for the 3rd millennium"

Bratislava, Slovakia, 30-31 May 2000

Resolution No 5/2000 on the draft Code for sustainability in sport: A partnership between sport and the environment

The European Ministers responsible for Sport, meeting in Bratislava for their 9th Conference from 30-31 May 2000

Convinced of the need for a partnership between sport and the environment which takes into account the existing international and national programmes directed towards sustainability in sport, together with those examples of good practice described in the Clearing House publication "Environment and Sport. Awareness raising, education, training",

Mindful that a clean and healthy environment is essential for those taking part in recreational or competitive sport and that many sports organisations and participants contribute actively to environmental protection,

Aware that the responsibility for sustainable development lies with all those active in sport and that the sports policy sector, the sports practice sector and the scientific sector, together with individual sportsmen and sportswomen, must be involved if progress is to be made in this field,

I Resolve:
to give their full support to the Code for Sustainability in Sport: A partnership between sport and the environment, reproduced in the Appendix,

II Invite the Committee of Ministers of the Council of Europe:
to adopt the "Code for Sustainability in sport: A partnership between sport and the environment" (see Appendix) as a Recommendation to Governments, to ask

the Committee for the Development of Sport (CDDS) to give appropriate follow-up to this Code, by setting up a working group,

III Resolve:
to distribute the Code, once adopted, in their own language, among national sports and environmental organisations and promote its dissemination to all the appropriate target groups, to encourage the authorities responsible for sport and the environment to work together to encourage regional, national and international sports organisations and federations to develop campaigns, literature and educational materials in this area.

Appendix

Draft code for Sustainability in Sport A partnership between sport and the environment

Aims and definition

This Code aims to set out guidelines for a working partnership between sport and the environment. It is important to ensure that sports can be enjoyed, both today and in the future, by as many people as possible in the most favourable conditions, by which is meant sport in a clean and healthy environment.

Sport in all its forms, practised at all levels, whether recreational or competitive sport, shares with the whole of human society the duty of nurturing and safeguarding the natural environment for both present and future generations. The adoption of the definition of sustainable development 'as development that meets the needs of the present without compromising the ability of future generations to meet theirs' is a starting point for action now.

The concept of sustainability in sport has already been incorporated into Article 10 of the European Sports Charter of 1992. This Article emphasises the responsibility of everyone involved in sport to protect the environment and promotes the introduction of a model of environmentally friendly sport. However, measures are needed to address those trends and practices in sport that are currently endangering the environment. The Code indicates realistic ways to ensure environmental protection. In practical terms this entails national policies that ensure respect for sustainability in sport in urban, open country and water areas.

The International Olympic Committee has organised conferences on sport and the environment, emphasised the need to respect environmental concerns when planning the Olympic Games and adopted Agenda 21, which reflects concern for the relationship between sport, the natural environment and sustainable development.

Responsibility for sustainable sport

This code focuses on the three sectors that bear the major responsibility for action which are:

- the policy sector, which is composed of groups and people concerned with the strategic choices, legislation, values and the underlying issues behind these questions. This includes governments and international organisations;
- the sports practice sector which is made up of sports organisations and federations, those engaged in the planning and construction of sports facilities, the educational sector, those concerned with sports tourism and the sporting goods industry. This group covers the whole spectrum from the large international and national sporting federations to the sports clubs at the grassroots level to the individual sportsman and woman;
- the scientific sector which embraces institutions, groups and individuals carrying out research, collecting and evaluating data and disseminating information.

Co-ordination between these sectors should be assured. Moreover the similarity of aims and concerns provides an opportunity to create close partnerships between those working for sustainable development in sport and those working for environmental conservation and protection. Environmental organisations can provide useful expertise.

Activities common to all three sectors

While each sector and its composite groups have a distinctive role to play in the area of sustainable sport there are areas where co-operation on common activities is essential. These include:

- exchanging information on progress and activities in their sector;

- emphasising the importance of education in environmental issues in relation to sport. Such education can take place in the framework of physical education and sports programmes in schools and universities, in sports clubs at all levels, in the programmes of international and national sports federations and associations and in sports management programmes;

- co-operating closely in the drawing up and implementation of international and national codes;

- alerting the media to the relevance and importance of this area and the need for closer co-operation. All three sectors should make a determined effort to involve the media in giving information on the need for measures and the actions being undertaken to halt the damage to the environment from sports activities;

- consulting on a broad basis to ascertain how widespread sports participation as favoured by 'Sport for All' can be made compatible with the environmental concerns for sports facilities, sports related traffic, access to the countryside and the control of the noise, waste and pollution which sport generates.

The policy sector
To guarantee the practice of sport in a way that will contribute to sustainable development, national, regional and local governments should, as appropriate to their respective competence:

a) examine the need for legislation and similar measures on:
- the use of non-renewable natural resources e.g. land, water, fossil fuels and the promotion of eco-friendly materials in the planning, designing, constructing, operating and maintaining sports facilities;
- transport provision in the siting of sports facilities, both indoor and outdoor open air sport, as a way of reducing sports related traffic and keeping sports participants and spectators away from sensitive areas and protecting fauna and flora;
- the control of forms of sport and associated technological practices and changes that carry potential harm for the environment;

b) study the possibility of introducing financial incentives:
- to help modernise sports facilities to reduce consumption of non-renewable resources;
- to subsidise the development of sports facilities and the holding of sports events in proportion to the concern shown for the environment in their planning, as is already often the case for the disabled;

c) review their sports policies from the viewpoint of environmental friendliness;

d) consult at the international level to ensure that national legislation is compatible with that of other countries. This would prevent a situation where practices are forbidden or regulated in one country but can be engaged in freely without restriction elsewhere;

e) support and encourage the introduction of a "green" label that could be awarded to products, facilities, events, organisations etc. which are taking environmental concerns seriously;

f) when hosting major sports events ascertain that environmental concerns are taken into account from the outset throughout the whole operation.

The sports practice sector

All the sports and sports related bodies (sports organisations and federations, those engaged in planning, designing, operating and constructing sports facilities, the educational sector, the sports media, those concerned with sports tourism and the sporting goods industry) should be involved in the drafting and implementation of policies and projects at the national level. They should take steps to remain abreast of the results and research coming from the scientific sector. These measures would indicate the desirability of mainstreaming environmental concerns into all their programmes and short and long term planning.

Possible measures include:
- the appointment of a person, at whatever level is appropriate to an organisation, with responsibility for environmental questions;
- the drawing up of curricula and provision of packs for environmental education at all levels;

- the involvement of well known national sportswomen and sportsmen to support sustainability in sport;
- liaison between the sporting goods industry and the sports organisations and federations to promote the use of products which respect environmental concerns, possibly by the institution of an award or a special label;
- active respect for environmental factors in the management of sports facilities;
- promotion of environmental sensitivity in sports tourism.

The scientific sector

This sector should co-operate closely with both the policy and sports practice sector to:

- undertake surveys and research to ascertain ways to limit damage to the environment through sport;
- investigate ways to measure environmental impact of sports, by for example evaluating a monitoring system of the costs and benefits of sport (for example sports events) in relation to potential damage resulting ;
- recommend new ways of sports involvement which can protect both the concept of "Sport for All" and promote sustainability in sport;
- collect and evaluate information on measures on both the international and national projects designed to encourage environmentally friendly sport;
- ensure the wide dissemination of information, research results and data on sport and sustainability to the sports community;
- provide advice and documentation to the various bodies that make up the policy and sports practice areas;
- analyse the various laws and codes for compatibility within Europe.

A common responsibility

Achieving sustainability in sport is a task for the whole world of sport. Taking active responsibility for the environment is a vital and important step towards the attainment of a clean and healthy sport in the 3rd millennium.

Everyone shares this responsibility and must ensure that their activities do not damage the environment, but rather safeguard and sustain it.

Literature

ABN (Arbeitskreis beruflicher und ehrenamtlicher Naturschutz), Hrsg.: Sport und Naturschutz im Konflikt. Jahrbuch für Naturschutz und Landschaftspflege, Bd. 38, 1986

ADAC Sport: Umweltplan 2000+, Umweltgerechter Motorsport auf dem Weg ins 21. Jahrhundert. München (2.Aufl.) 1997

ADFF (Arbeitsgemeinschaft der Deutschen Fischereiverwaltungsbeamten und Fischereiwissenschaftler) (Hrsg.): Fischerei in Naturschutzgebeiten. H. 6 der Schriftenreihe der ADFF, 2. Aufl., Offenbach am Main, 1993

AFTSC (Association de Fédérations de Tir Sportif de la C.E.E.): "Blei und das Wurftaubenschießen" und "Lärmbelästigung durch Sportschießen"? Informationsblätter Nr. 1 und Nr. 2, Paris (ohne Jahresangabe)

ALBRECHT, R. / PAKER L. / REHBERG S. / REINER, Y.: Umweltentlastung durch ökologische Bau- und Siedlungsweisen, Bd. 1: Planungsvorschläge und bauliche Maßnahmen, Bd. 2: Auswirkungen und Baustoffverwendung, Energiebedarf, Luft und Klima, Abfallbeseitigung und Wasserhaushalt, Lärm, Flächenbedarf, Kosten und Arbeitsmarkt. Bauverlag Wiesbaden/Berlin 1984

AMMER, U. / TIETZE, H.: Reiten in der offenen Landschaft - Eine Studie zur Verdeutlichung der Konfliktsituation und ein Beitrag zur Problemlösung. In: Forstw. Chl. 98, H.4, 1979

ANL (Akademie f. Naturschutz und Landschaftspflege): Fischerei und Naturschutz, Tagungsbericht 1981

BARTH, H.J.: Loipen- und Wegekonzept für den Inneren Bayerischen Wald mit St. Englmar und Wegscheid auf der Basis naturschutzfachlicher Grundlagen. In.: Schriftenreihe "Naturschutz und Landschaftspflege", Bd. 131, Bayer. LfU (Hrsg.), München 1995

BAUER, H.: Motorsport und Umwelt - Darstellung eines problembeladenen Beziehungsgefüges unter besonderer Berücksichtigung von Lösungsansätzen, Europäische Hochschulschriften Peter Lang 1989

Bay LFU (Bayer. Landesamt f. Umweltschutz) und ANL (Akademie f. Naturschutz u. Landschaftspflege): Naturschutz und Golfsport. Merkblätter zur Landschaftspflege und zum Naturschutz 2, München 1989

BayLfU (Bayer. Landesamt f. Umweltschutz Hrsg.): Erholungslenkung für Skilanglauf und Wandern, Seminarbericht 22. Nov. 1995

BAYSTMLU (Bayer. Umweltministerium): Umweltgerechte Bootshäfen - eine Leitlinie für Hafenbetreiber. Fachinformation "Umwelt und Entwicklung" 1/1994

BAYSTMLU (Bayer. Umweltministerium): Der umweltbewußte Sportverein, Leitfaden, München 1996

BICHELMEIER, F.: Klettern-Naturschutz ein Konflikt? In: Berichte des Bayer. Landesamtes f. Umweltschutz (4), Schriftenreihe H. 108, München 1991

BISP (Bundesinstitut f. Sportwissenschaft): Planung, Bau und Unterhaltung von Golfplätzen. Schriftenreihe Sport und Freizeitanlagen, Planungsgrundlagen P 1, Köln 1987

BISP (Hrsg.): Die Sportanlagenlärmschutzverordnung und ihre Auswirkungen in der Praxis. Schriftenreihe Sportanlagen und Sportgeräte, Informationen I 1/94

BISP (Hrsg.): Grundsätze zur funktions- und umweltgerechten Pflege von Rasensportflächen: Nährstoffversorgung durch Düngung (Teil I, 1993), Wassersparende Maßnahmen (Teil II, 1994), Unerwünschte Pflanzenarten auf Rasensportflächen (Teil III, 1995), Pflanzenkrankheiten und Schädlinge (Teil IV, 1996), Köln

BISP: Grundsätze zur funktions- und umweltgerechten Pflege von Rasenflächen, Teil I "Nährstoffversorgung durch Düngung" (1993) und Teil II "Wassersparende Maßnahmen" (1994), Köln 1993/94

BML (Bundesministerium für Ernährung, Landwirtschaft und Forsten, Hrsg.): Umweltverträgliche Reitwegeplanung - Modell Rheinisch-Bergischer Kreis. (Bearb. Fleischhauer, K. / G. Ruwenstroth, B. Winkler, H. Strodthoff, GfL Bremen) Bonn 1984

BRAHM, G.: Ein Weg zum Erhalt des Klettersports im Nördlichen Frankenjura. In: DAV-Naturzschutz-Info 2, 1993

BRÄMER, R.: Wandern - der sanfte Natursport. In: Die Eifel. H. 1, 1997

BRESSLER, G. / GÜNTHER-POMHOFF, C.: Analyse und Energiekonzept einer Berggaststätte. In: Energiewirtschaftliche Tagesfragen. 47. Jg., H. 5, 1997

BRÜMMER, F. / PÜTSCH, M.: Leitbilder eines natur- und landschaftsverträglichen Tauchens. In: Vorlagen zum Kongreß "Leitbilder eines natur- und landschaftsverträglichen Sports" in Wiesbaden. Hrsg. Deutscher Naturschutzring, 1996

CERNUSCA, A., ANGERER, H., NEWESELY, Ch., TAPPEINER, U.: Auswirkungen von Schneekanonen auf alpine Ökosysteme. Ergebnisse eines internationalen Forschungsprojektes. In: Gnaiger, E. & Kautzki, J. (Hrsg.): Umwelt und Tourismus, Wien 1992

CERNUSCA, A.: Probleme von Wintersportkonzentrationen für den Naturschutz. In: Jahrbuch für Naturschutz und Landschaftspflege, ABN (Hrsg.) a.a.O., Bd. 38, 1986

CONRAD / BRENNER: Naturschutzgebiet Monheimer Baggersee - Kompromißvorschlag für die räumlich zeitliche Trennung von Angelsport und Artenschutz unter besonderer Berücksichtigung des Angelns vom Boot aus. Gutachten der LÖF und der Landesanstalt für Fischerei NW (Hrsg.), Recklinghausen/Albaum 1986

CURIO, E.: Proximate on Developmental Aspects of Antipredator Behavior. Advances in the Study of Behavior. Vol. 22, Pages 135-238, Academic Press, Inc., 1993

DA COSTA, L.P. (Ed.): Environment and Sport - An International Overview. Faculty of Sport Sciences and Physical Education, University of Porto, 1997

DAEC (Deutscher Aero Club): Verhaltenskodex der Luftsportler - Für umweltbewußten Luftsport im Deutschen Aero Club, Heusenstamm, 1997

DAV (Deutscher Alpenverein): "Sanft Klettern" - der Natur zuliebe. Merkblatt für Kletterer, München 1987

DAV: Entsorgungsstudie für Alpenvereinshütten, Teil A: Abwasser, München 1989

DAV: Zu Gast in den Felsen. Merkblatt zum naturschonenden Klettern. München o.J.

DAV: Alpenvereinshütten zapfen Energie von der Sonne (Broschüre), München o.J.

DAV: Erleben mit gutem Gefühl. Informationen für Skibergsteiger, Tiefschneefahrer und Snowboarder. Faltblatt, München o.J.

DAV (Deutscher Anglerverein): Pro Natur. Dresden 1995

DAV: Positionspapier des Deutschen Angelverbandes zum Schutz von Natur und Umwelt. Berlin-Lichtenberg 1996

DEUTSCHE REITERLICHE VEREINIGUNG (Hrsg.): Umweltschutz und Tierschutz. In: Handbuch für Reit- und Fahrvereine, Loseblattsammlung, Warendorf 1989

DEUTSCHE SCHÜTZENJUGEND: Sportschützen nehmen Rücksicht auf die Natur. In: Info-Blatt Nr. 22, 1989

DICKMANN, J.: Schäden durch Reiten im Wald. In: Natur und Landschaft, H.5, 1985

DJH (Deutsches Jugendherbergswerk): Handbuch Umweltschutz und Umwelterziehung in Jugendherbergen, Ringbuch, Detmold 1990

DKV (Deutscher Kanuverband) (Hrsg.): Natur- und Gewässerschutz. Schriftenreihe des DKV, Bd. 6, Duisburg 1986

DKV (Hrsg.): Kanuwandern und Naturschutz - Wege zum naturbewußten Paddeln. Schriftenreihe des DKV, Bd. 10, Duisburg 1997

DMYV: Die Möwe Emma empfiehlt. Umweltschutz-Ratschläge für die Sportschiffahrt (Faltblatt)

DNR (Deutscher Naturschutzring) Hrsg.: Leitbilder eines natur- und landschaftsverträglichen Sports, Vorlagen zum Kongreß. Wiesbaden 1996

DSV (Deutscher Seglerverband): Energie aus Sonne, Wind und Wasser an Bord und an Land, Hamburg 1990

DSV: Bau- und Betriebsempfehlungen für umweltgerechte Sportboothäfen, Hamburg 1993

EAG (Europäischer Golf Verband): Europäische Umweltarbeitsgruppe. In: Sport schützt Umwelt, H. 37, 1995

EIRICH, R. / ROSKAM, F. / SKIRDE, W. / PÄTZOLD, H.: Sportplatzbau und -unterhaltung, Deutscher Fußballbund (Hrsg.), Frankfurt 1989

ELSASSER, H.: Durch den Wandertourismus verursachte Schäden, dargestellt am Tourismus im Schweizerischen Nationalpark. In: IGU (Internationale Gesellschaft für Umweltschutz) (Hrsg.) Tagungsband ENVIROTOUR, Wien 1992

ENGELHARDT, W.: Die umweltverträgliche Reitanlage. In: Reitsport 2000 (Deutsche Reiterl. V., Hrsg.), Kongreßbericht Equitana, Warendorf 1987

FICHT, B. / HEPP, K. / KÜNKELE, G. / SCHILLING, F. / SCHMID, F.: Lebensraum Fels. In: Beih. Veröff. Naturschutz Landschaftspflege Baden-Württemberg, Bd. 82, Karlsruhe 1995

GEORGII, B./ SCHRÖDER, W. / SCHREIBER, R.L.: Skilanglauf und Wildtiere, Konflikte und Lösungsmöglichkeiten. Schriftenreihe ökologisch orientierter Tourismus, Alpirsbach 1984

GOLD, R. / KNEBEL, W. / PUTZER, D.: Mauserplätze für bedrohte Wasservogelarten - Planungskonflikte und Erfahrungen mit Freizeitnutzungen im Rheinland. In: Naturschutz u. Landschaftsplanung, H. 4, 1993

HAASS, H. (Hrsg.): Handlungsrahmen zur Standortplanung von Wassersportanlagen im Spannungsfeld von Nutzerattraktivität, Ökologie und Ökonomie. Reihe Sportwissenschaft Bd. 1, Münster 1996

HABER, W.: Zur landschaftsökologischen Beurteilung von Golfplätzen. In: Golfmagazin, H. 3, 1983

HEINZEL, R. / ZIMMERMANN, M.: Handbuch Umweltschonende Großveranstaltungen - Leitfaden für Planung und Durchführung unterschiedlicher Veranstaltungstypen. Berlin 1990

HOFFMANN, M. / POHL, W.: Felsklettern, Sportklettern. Alpin-Lehrplan Bd. 2, Hrsg. Alpenvereine, BLV-Verl. München 1996

INTERNATIONALE GEWÄSSERSCHUTZKOMMISSION für den Bodensee: Limnologische Auswirkungen der Schiffahrt auf dem Bodensee. Bericht Nr. 29 (1982) und Nr. 31 (Schadstoffe in Bodensee-Sedimenten), 1984

JÄGEMANN, H.: Perspektiven eines dauerhaft umweltverträglichen Sports. Referat (unveröff.) im Rahmen einer Vortragsreihe der Universität Hannover am 26.6.1996

JÄGEMANN, H. / STROJEK, R. (Hrsg.): Fließgewässer und Freizeitsport - Dokumentation der Fachtagung "Ökologische Bewertung von Sport- und Freizeitaktivitäten an Fließgewässern, Frankfurt 1996

JOB, H.: Tourismus versus Naturschutz: "sanfte" Besucherlenkung in (Nah-) Erholungsgebieten. In: Naturschutz und Landschaftsplanung, H. 1, 1991

KAYSER, C.: Auswirkungen von wasserorientierten Sportaktivitäten auf Fließgewässerökosysteme. Dipl.-Arb. Univ. Hannover 1994

KELLER, T. / VORDERMEIER, T. / LUKOWICZ, M. / KLEIN, M.: Der Einfluß des Kormorans auf die Fischbestände ausgewählter bayerischer Gewässer. In: Fischer und Teichwirt, Jg. 47, H. 3, 1996

KIRST, C.: Flugsportanlagen in der Bundesrepublik Deutschland und ihr Konflikt mit dem Naturschutz. In: Natur und Landschaft, H. 7/8, 1989

KOEPFF, C. / DIETRICH, K.: Störungen von Küstenvögeln durch Wasserfahrzeuge. In: Die Vogelwarte, Jg. 33, 1986

KUHN, P.: Indoor-Sport und Ökologie. Beiträge zu Lehre u. Forschung im Sport Bd. 113, Schorndorf 1996

LAUTERWASSER, E.: Skisport und Umwelt - Ein Leitfaden zu den Auswirkungen des Skisports auf Natur und Landschaft. DSV-Umweltreihe Bd. 1, Weilheim 1989

LAUTERWASSER, E.: Integrales Modellprojekt Rohrhardsberg/Martinskapelle. In: Sport schützt Umwelt, Nr. 23, 1991

LAUTERWASSER, E./ ROTH, R.: Spurenwechsel zum umweltbewußten Skisport. DSV-Umweltreihe Bd. 5, Weilheim 1995

LEICHT, H.: Beschneiungsanlagen und Naturschutz - eine naturschutzfachliche Betrachtung der Situation in Bayern. In: Natur und Landschaft, H.2, 1993

LEICHT, H. / T. DIETMANN / U. KOHLER: Landschaftsökologische Untersuchungen in den bayerischen Skigebieten. In: Naturschutz und Landschaftsplanung, H. 3, 1993

LEUMANN, P.: Umweltverträgliche Bodenbeläge und umweltfreundliche Anlagenpflege. In: IAKS (Hrsg.) Internationaler Kongreß Sportstättenbau und Bäderanlagen des IAKS, Köln 1985

LIPSKY, H.: Auswirkungen des Skilanglaufens auf Vegetation und Flora von Feuchtgebieten. Gutachten im Auftrag des Bayer. LfU, München 1996

MINISTERIUM F. UMWELT Baden-Württemberg (Hrsg.): Leitfaden zur landschaftsbezogenen Beurteilung und Planung von Golfanlagen LfU, Karlsruhe 1989

MOSIMANN, T.: Beschneiungsanlagen in der Schweiz - aktueller Stand und Trends - Umwelteinflüsse - Empfehlungen für Bau, Betrieb und UVP. Geosynthesis 2, Universität Hannover und Universität Bern 1991

MOSIMANN, T. / LUDER, P.: Landschaftsökologischer Einfluß von Anlagen für den Massenskisport. Materialien zur Physiogeographie H. 1, Basel 1980

NABU (Naturschutzbund Deutschland) (Hrsg.): Überbelichtet - Vorschläge für eine umweltfreundliche Außenbeleuchtung. Kornwestheim 1994

NIESL, G. / PROBST W. / HINGSAMER, J.: Die Geräuschemission von Tennisanlagen. Zeitschrift für Lärmbekämpfung 30, 1983

NOEKE, J / ROLF, A.: Maßnahmen zur Lärmsanierung und Lärmvorsorge auf wohnnahen Freizeitanlagen. INFU Werkstattreihe H. 17, Univ. Dortmund 1986

OMK (Oberste Motorradsport-Kommission), Hrsg.: Umwelt Code der FIM, Frankfurt 1997

ONS (Oberste nationale Sportkommission für den Automobilsport in Deutschland): Automobil-Breitensport: Einstieg leicht gemacht! 2. Aufl. Frankfurt 1993

PÄTZOLD, H.: Lärmschutzeinrichtungen im Sportplatzbau - Wirkung und Ausführung. In: BISP 1994 a.a.O.

PUTZER, D.: Segelsport vertreibt Wasservögel von Brut-, Rast- und Futterplätzen - Störung durch Boote geländeökologisch und mathematisch erfaßt. In: Mitteilungen der LÖLF, H. 2, 1983

PUTZER, D.: Angelsport und Wasservogelschutz in Nordrhein-Westfalen - Welchen Raum läßt der ordnungsgemäße Angelsport dem Artenschutz? Analysen, Fragen und Antworten. In: Ber. Dtsch. Sekt. Int. Rat Vogelschutz 25 (1985) S. 65-76

RANFTL, H.: Auswirkungen des Luftsports auf die Vogelwelt und die sich daraus ergebenden Forderungen. In: ANL Berichte 12, 1988

REBHAN, H. (1992): Besiedlung oberfränkischer Flugplätze und ausgesuchter Vergleichsflächen. In: Berichte der Bayerischen Akademie f. Naturschutz u. Landschaftspflege ANL Bd. 16, S. 215-227

REICHHOLF, J.: Auswirkungen des Angelns auf die Brutbestände von Wasservögeln im Feuchtgebiet von internationaler Bedeutung "Unterer Inn". Vogelwelt Jg. 109, S. 206-221, 1988

REICHHOLF, J. / SCHEMEL, H.J.: Segelsport und Naturschutz - gehört das Segeln aus ökologischer Sicht zur "ruhigen Erholung"? in: Zeitschrift für angewandte Umweltforschung, H. 4, 1988

ROSSBACHER, R.: Vogelschutz und Modellflugsport. In: Vogel und Umwelt, H. 1, 1982

SCHAUER, TH.: Vegetationsveränderungen und Florenverlust auf Skipisten in den bayerischen Alpen. Jahrbuch Verein z. Schutz der Bergwelt, 46. Jg., München 1981

SCHEMEL, H.J.: Umweltverträgliche Freizeitanlagen - eine Anleitung zur Prüfung von Projekten des Ski-, Wasser- und Golfsports aus der Sicht der Umwelt, Bd. 1 (Analyse und Bewertung), Berichte des Umweltbundesamtes, Bd. 5, 1987

SCHEMEL, H.J.: Naturerfahrungsräume - Ein humanökologischer Ansatz zur Sicherung von naturnaher Erholung in Stadt und Landschaft. Angewandte Landschaftsökologie, Heft 19, Bonn-Bad Godesberg, 1998

SCHEMEL, H.J. / ERBGUTH, W.: Handbuch Sport und Umwelt - Ziele, Analysen, Bewertungen, Lösungsansätze, Rechtsfragen. Langfassung. Aachen 2000 (3. Auflage)

SCHEMEL, H.J. / STRASDAS, W.: Bewegungsraum Stadt - Bausteine zur Schaffung umweltfreundlicher Sport- und Spielgelegenheiten. Aachen 1998

SCHEUERMANN, M.: DAV-Projekt "Skibergsteigen umweltfreundlich" im Rahmen der Untersuchung "Skilauf und Wildtiere" des Bayer. Umweltministeriums. In.: DAV-Mitteilungen, H. 2 und H. 4, 1996

SCHNAUBER, H. / SCHMIDT, K.: Lärmimmissionen und Möglichkeiten vorbeugender Maßnahmen - Ein Leitfaden des Deutschen Tennis Bundes, Göttingen o.J.

SCHNEIDER-JACOBY, M / BAUER, H.-G. / SCHULZE, W. (1993): Untersuchungen über den Einfluß von Störungen auf den Wasservogelbestand des Gnadensees (Untersee/Bodensee). Ornithologische Jahreshefte für Baden Württemberg, 9 (1), 1-24

SCHRÖDER, W. / DIETZEN, W. / GLÄNZER, U.: Das Birkhuhn in Bayern. Schriftenreihe Naturschutz u. Landschaftspflege des Bayer. Landesamtes f. Umweltschutz, H. 13, München 1981

SCHULTHEISS, H.: Motocross: Gefahren für Natur und Landschaft. In: ABN (Hrsg.) Sport und Naturschutz im Konflikt a.a.O., 1986

SENN, G.-T.: Klettern und Naturschutz. Bd. 4 der Reihe "Mensch-Natur-Bewegung", Hrsg. R. Strojec, Rüsselsheim 1995

SKIRDE, W.: Untersuchungsergebnisse zur Belastbarkeit von Rasennarben durch Maßnahmen der Sportplatzpflege. Berichte B1/88 Schriftenreihe Sport- und Freizeitanlagen des Bundesinstituts für Sportwissenschaft, Köln 1988

STB (Schwäbischer Turnerbund): STB-Tips "Sport und Umwelt", Das Know-how für die Organisation umweltverträglicher Sportveranstaltungen, Stuttgart 1993

STEIN, H.: Ursachen der Bestandsgefährdung: Fischereiliche Maßnahmen. Schriftenreihe der Arbeitsgemeinschaft der Deutschen Fischereiverwaltungsbeamten und Fischereiwissenschaftler, H.1/86: Fischerei und Fischartenschutz, 1986

STIBBE, A.: Sporttauchen (mit Kapitel "Tauchen und Umwelt") VDST (Hrsg.), 7. Auflage, Stuttgart, 1994

TWIEHAUS, K.: Mountainbikefahren und Umwelt im Gebirge - Einflüsse, Schäden und Konflikte. Unveröffentlichte Diplomarbeit am Geographischen Institut der Universität Hannover, 1994

UBA (Umweltbundesamt) Hrsg.: Die Belastung von Böden auf Sportschießplätzen durch Bleischrot und Wurftauben. UBA-Texte 39/89, Berlin 1989

UBA (Umweltbundesamt) (Hrsg.): Kriterien des Bodenschutzes bei der Ver- und Entsiegelung von Böden. Berlin 1994

UBA (Hrsg.): Leitfaden zum ökologisch orientierten Bauen, 3. Aufl., Heidelberg 1997

UBA (Hrsg.): Umweltfreundliche Beschaffung - Handbuch zur Berücksichtigung des Umweltschutzes in der öffentlichen Verwaltung und im Einkauf, 4. Aufl. Berlin 1997

UMWELTBUNDESAMT: Lärmbekämpfung '88, Tendenzen - Probleme - Lösungen, Berlin 1989

UPPENBRINK, M. / BOYE, P.: Strategiekonzepte zum Schutz der Wasservögel im Ostseeraum aus der Sicht einer nationalen Naturschutzbehörde. In: Natur und Landschaft, H. 9, 1995

VDGW: Leitlinien des Verbandes Dt. Geb. u. Wanderver. e.V., verabschiedet in der Mitgliederversammlung am 26. Juli 1996 in Wernigerode, Saarbrücken 1996

VOLK, H.: Skilanglauf und Umwelt, Ansichten und Verhaltensweisen der Läufer. In: Allgemeine Forst Zeitschrift (AFZ), H. 12, 1988

WAGNER, D.: Pferdesport und Umweltschutz. In: Die Deutsche Sportjugend und der Umweltschutz, Frankfurt 1984

WEISS, H.: Auswirkungen von Skianlagen und Geländekorrekturen aus der Sicht des Landschaftsschutzes. In: Bossard (Hrsg.): Skipistenplanierungen und Geländekorrekturen. Berichte Nr. 237 der Eidgen. Anstalt f. d. forstliche Versuchswesen, Birmendorf 1982

WILKEN / NEUERBURG: Umweltschutz im Sportverein, ein Ratgeber. Aachen 1997

WITTY, S. / KÖHLER, S.: Seit der Eiszeit überlebt - heute bedroht? Schutzkonzept des DAV für außeralpine Felspflanzen. In: Praxis der Naturwissenschaften - Biologie, H. 2, 45. Jg., 1996

WÖHRSTEIN, T.: Ökologische Auswirkungen des Mountainbike-Sports. Unveröffentlichte Diplomarbeit im Fachbereich Geographie der Universität des Saarlandes, 1993

WOLF, G.: Schutz und Pflege von Biotopen und Golfplätzen. In: Garten und Landschaft, H. 6, 1986

WYSER, M. / JORDI, B.: Die Abgasvorschriften für Motorschiffe weisen den Weg zur Schadstoff-Reduktion von Offroad-Motoren. In: BUWAL-Bulletin 3/97

XYLANDER, W.: Einflüsse des Tauchsports auf die biologische Umwelt. In: Lüchtenberg, D. (Hrsg.) Tauchen an Schulen und Hochschulen. Verlag S. Naglschmid, Stuttgart, 1991

ZIESE, J. / WULFERT, G.: Junge Disziplinen des Luftsports und ihre Auswirkungen auf die Natur. In: LÖLF-Mitteilungen, H. 1, 1989

ZUMKOWSKI, H. / XYLANDER, W.: Auswirkungen des Tauchsports, insbesondere der Anfängerausbildung, auf Seen und Meere. In: Anfängerausbildung im Tauchsport (Uwe Hofmann, Hrsg.), MTi Press, Stuttgart. Divemaster Workshop 1, 1995